No. 689
$7.95

Cassette Tape Recorders
How They Work — Care & Repair

By Walter G. Salm

TAB BOOKS
Blue Ridge Summit, Pa. 17214

FIRST EDITION

FIRST PRINTING— OCTOBER 1973

Printed in the United States
of America

Hardbound Edition: International Standard Book No. 0-8306-3689-7

Paperbound Edition: International Standard Book No. 0-8306-2689-1

Library of Congress Card Number: 73-86291

PREFACE

In the few short years since the Philips cassette was introduced to the American market at the 1964 High Fidelity Show, this ingenious tape format has grown in importance and sales until it is now the number one selling tape medium in the world. Today, cassettes are not only used for their convenience in handy battery-powered portable machines, they are also used for compatible stereophonic music systems, both in the living room and on the highway, and are even finding their way into office machines and computers.

Recording tape has always had the potential for being a vast and ever-present influence in our personal and business lives. But it was with the advent of the cassette that many of tape's discouraging characteristics—threading, tape spill, bulky reels, and relatively large-size equipment—were eliminated. The Philips cassette has changed this picture, drastically. This book is a case in point; 19 chapters devoted only to cassettes—something that would have been unthinkable only a couple of years ago.

When N.V. Philips Gloeilampenfabrieken introduced the cassette format, it was immediately envied and much copied by other manufacturers. Following the Dutch company's lead, manufacturing giants in Europe and in the Far East started introducing their own cassettes and equipment for them. The problem, then, was a lack of standardization; no one manufacturer's cassette would fit anyone else's machine.

Since Philips held definitive patents for the compact cassette, and probably because of the company's enormous resources, it held a trump card. Soon, Philips was offering to license any manufacturer in the world to make its cassettes and equipment for them on a first-come, first-served basis. The only stipulation was that such manufacturers sign an agreement to adhere strictly to the standards and formats that would be dictated by Philips.

The success of this move was immediately apparent; within a couple of months nearly 60 companies in every major manufacturing country in the world had signed up, and today the licensees number in the hundreds. There is no longer such a thing as a Grundig cassette, or an Aiwa cassette; those were among the systems that were born in the wake of the first cassette announcement and faded from existence soon afterwards. Today, all cassettes adhere to the Philips format, whether they are made in Holland, the United States, Japan, Germany, Mexico, or anywhere else in the world.

To be sure, there are slight variations—some companies like to put metal-foil sensing strips at the ends of the tapes for tripping particular types of mechanisms. Other companies vie with one another for the claim of the most superior mechanical characteristics of the cassette package. And there is an oxide war raging among major tape manufacturers with each one making fantastic claims for his own product.

We do not attempt to make value judgments about the various systems offered; instead, we explain how they work, how they differ, and what may be expected of them. This philosophy holds true in our discussions of both the tape and equipment. We endeavor to present a well-rounded picture of the cassette marketplace and what is available. We show principal applications for cassettes and typical kinds of equipment that are suited for these uses. We also cover various types of installations for a variety of equipment and show some of the more important steps in maintaining both equipment and tapes in topnotch running order. We also cover some basic repairs and the best way to handle them in the home and in the shop.

In effect, this is a cassette handbook and should be treated as such. We believe firmly in the cassette—in fact, we're downright prejudiced in its favor as opposed to the continuous-loop cartridge that is so popular for automotive and living-room use today. This prejudice is based on the evidence of years of experience with both cassettes and cartridges and personal preference for systems that are relatively trouble free.

Walter G. Salm

CONTENTS

THE CASSETTE TAPE MEDIUM

Chapter 1

About 15 years or so ago, someone at RCA had a bright idea. He took that cumbersome open reel of tape that was being peddled for the home market and put it inside a sealed plastic box (Fig. 1-1). It seemed like a great idea; it eliminated troublesome tape threading and provided instant loading and unloading of equipment with no rewinding necessary.

So RCA went ahead full tilt, making its tape cartridges and equipment. Other manufacturers liked the idea also, and several companies started to manufacture equipment to handle the RCA-type cartridges. Significantly, some of these companies are no longer in business.

The RCA idea had merit, and if it hadn't been for the cumbersome size of the cartridge, it might have survived to this day. It did manage to hang on for several years, but finally gave way to more far-reaching developments.

ORIGIN OF THE CASETTE

The Philips cassette is very much like the ill-fated RCA cartridge in many respects, except for one important factor—size. The Philips cassette is perhaps one-sixteenth the size of the old RCA cartridge, and less than one-quarter the size of today's 8-track stereo cartridge, the type that is so popular among automotive listeners.

The compact cassette has few if any of the disadvantages of the cartridge, and it has many peculiar advantages of its own. One thing is size. The cassette measures a scant 2 1 2 x 4 inches and is about 7/16 inch thick at its thickest point. The tape is a fine ribbon of plastic measuring barely 1/7 inch in width and can be crammed with four tracks of stereo information.

Fig. 1-1. Remote ancestor of the cassette, the RCA cartridge used ¼-inch tape, had many of the cassette's features, but was ungainly in size.

Basic Cassette

The basic cassette has three square openings to permit access to the tape which passes through these holes with the oxide facing out (see Fig. 1-2). The business end of the cassette also has a series of four round holes. Two of them are used in the recording and playback process. The first round hole accepts a needle-thin capstan; the square cutout in front of it admits the rubber pinch roller which presses the tape against the capstan shaft.

The second and third round holes are used for alignment purposes, bellying up to a couple of short alignment pins in the

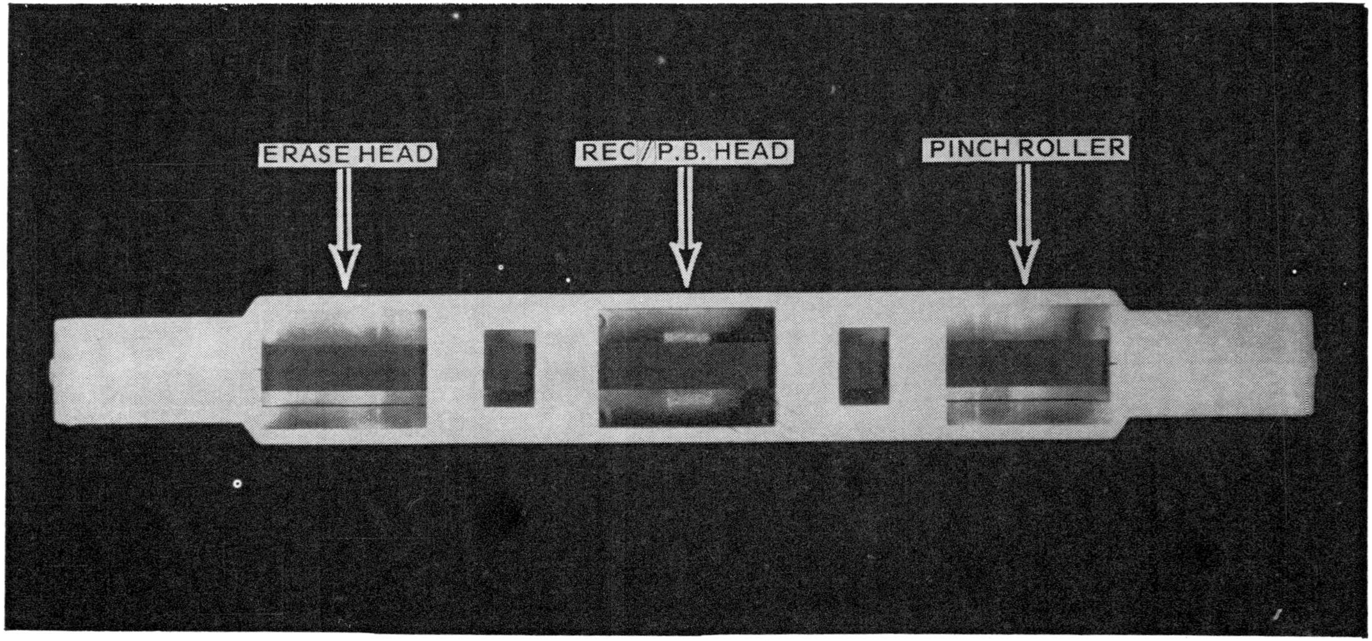

Fig. 1-2. The Philips-type cassette has three square holes: the left for the **erase head**, center for the record-playback head, and the right for the rubber pinch roller. When inverted for side 2 use, holes one and three **reverse** roles. (Courtesy TDK.)

Fig. 1-3. The cassette's window shows how much tape is left on feed hub, how much has been used. With practice, it's possible to gage by eye how much playing time is left.

recorder. The middle square hole has a felt pressure pad behind the tape, and this is where the recording and playback head presses against the oxide. The third square hole is for the erase head. When the cassette is turned over, the first and third holes reverse roles. That was the original concept and is still true in about 90 percent of all machines in use.

But those holes have also proved to be a serious limitation to some manufacturers, and a great deal of ingenuity has gone into engineering equipment that could make better use of these limited apertures in more sophisticated equipment. Machines of this type, especially those with automatic reversing, are covered in later chapters.

The cassette contains a rectangular window on each side, which permits the tape supply to be viewed at all times (see Fig. 1-3). Basically, the cassette encloses a tiny reel-to-reel tape, with the tape ends firmly locked into the takeup and feed hubs. The cassette is designed so that tape winding onto the takeup hub uses space that has just been vacated by tape unwinding from the feed hub. This makes for tight quarters inside the cassette, and the marvel of it all is that it works.

Antierase Feature

Another ingenious feature designed into the Philips cassette is a special antierase feature. The cassette provides a virtually foolproof method that prevents accidental erasure of valuable tapes that you would like to keep. The rear end of each cassette has two removable plastic tabs that can be pried out easily with the point of a penknife. One tab is for each direction of tape travel. With the tab removed (see Fig. 1-4), you cannot record on that side of the tape, since a built-in interlock finger (see Fig. 1-5) in the recorder will not be depressed. It's a relatively simple system and yet it represents the height of ingenuity.

The antierase interlock is incredibly simple. It consists of a metal finger in the cassette recorder—in all cassette recorders—which must be pressed backwards by the presence of the plastic tab before the red record button is released. If this interlock is not pushed, it's physically impossible to depress the record function button (see Fig. 1-6).

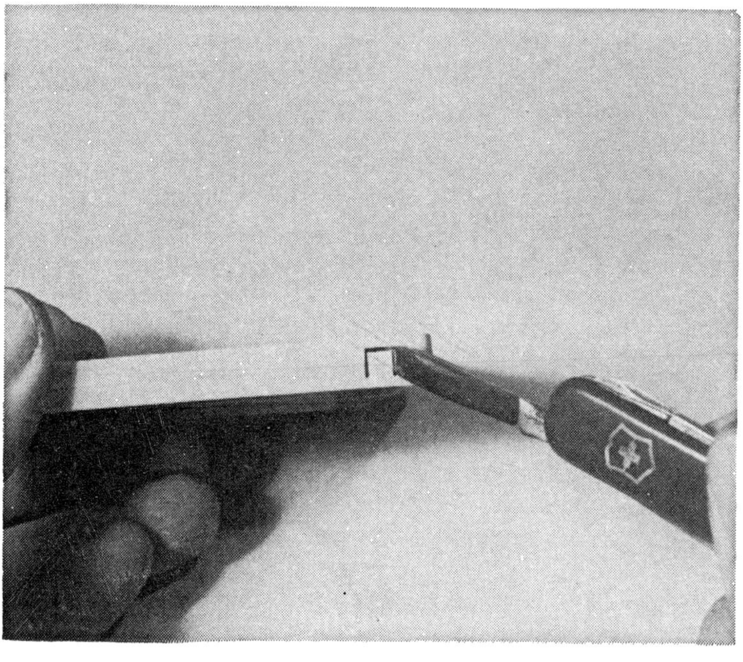

Fig. 1-4. To preserve valuable material from accidental erasure, simply pry out and break off the tab on the cassette's rear. One tab is provided for side 1, another tab for side 2.

Fig. 1-5. A record interlock finger is standard in all cassette recorders. The finger must be pressed to the rear of the cassette compartment before the red record button can be depressed.

When the time comes that one of those important tapes can be reused, all you have to do is put a piece of adhesive tape over the open hole in the back of the cassette. This will be enough to push back the interlock finger.

MONO AND STEREO FORMATS

Cassettes today can be used in one of three different formats, and all of them are totally compatible with each other. Unlike open-reel tape, a stereo recording on a cassette is interchangeable with a monophonic one. Mono cassettes can be played back on stereo machines, stereo cassettes can be played back on mono machines. The reason for this is apparent in the track layout shown in Fig. 1-7.

The monophonic cassette uses one-half of the tape width for one direction of travel and the other half for the other direction, the same as any half-track mono tape. At the end of the first side, the cassette is simply flipped over to use the other half, just as open-reel tapes are flipped over for mono half-track recording and playback.

When standards were set by Philips for stereophonic cassette recording, the track arrangement shown in Fig. 1-7 placed left and right channels side by side. You may recall that the 4-track stereo format for open-reel tapes separates left and right channels from each other, alternating with left and right for the other direction of tape travel. This was necessary because at the time that these standards were adopted by the tape industry, there simply was no way of putting left and right channels next to each other. The technology for manufacturing recording and playback heads just wasn't advanced enough then.

By the time standards were set for stereo cassettes in the late 1960s, magnetic and ferrite technology had been so well developed that the side-by-side arrangement was feasible. If a stereo cassette is played on a mono machine, the half-track monophonic head covers both left and right channels, providing a fully balanced single-channel program.

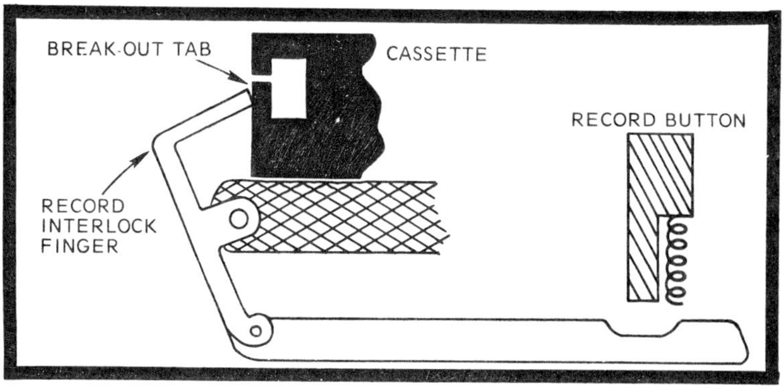

Fig. 1-6. Mechanical concept of how the interlock finger works with anti-erase breakout tab on rear of cassette. When the tab is removed, the record button cannot be depressed.

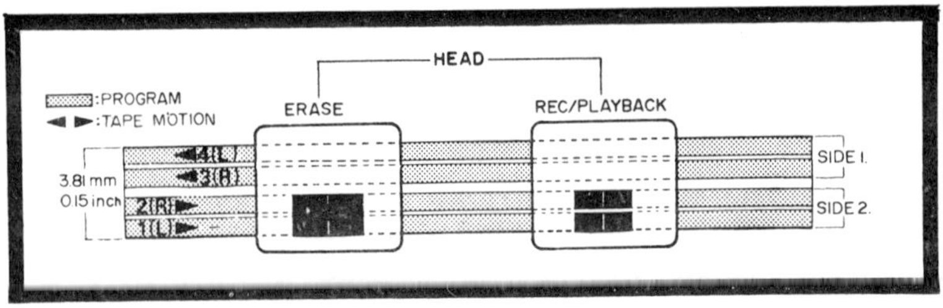

Fig. 1-7. Half-track mono and 4-track stereo track configuration for cassette tapes. Unlike open-reel tapes, left and right channels for the same direction of tape travel are side-by-side, providing monophonic playback compatibility. (Courtesy TEAC.)

The third format involved 4-channel stereo, or "quad stereo," as it is often called. Since the cassette is rapidly becoming the world's most popular medium, it stands to reason that a format for this new "surround sound" should be established for the cassette. In fact, several different companies have already set up equipment for using the cassette to provide four separate, discrete channels of audio. However, N. V. Philips waited until its own engineers had developed a system which is once again fully compatible with all other cassette formats. The difference between the Philips 4-channel stereo and systems developed by other companies such as Astrocom/Marlux is as shown in Fig. 1-8.

The Astrocom/Marlux system is not totally compatible with existing systems. The cassette can be played in one direction only on a regular stereo machine for ordinary left and right channels. The Philips system, instead, divides the tracks in half so that the very narrow 1/7-inch wide tape carries a total of eight tracks!

The reasoning is that when played on an ordinary stereophonic system, a full stereo program will be heard by the listener. Yet, the system provides the four channels of sound demanded in quad stereo systems with the same tape economy of two-channel stereo. The Philips system lets you turn the cassette over for side B playback; the Astrocom/Marlux system does not.

There are many arguments pro and con, and so far the industry has been very reluctant to accept quad stereo in cassette form. The advantages of the Philips system are apparent; its biggest disadvantage is that signal strength is greatly reduced by the narrow track width, and this means a serious downgrading of signal-to-noise ratio. The non-

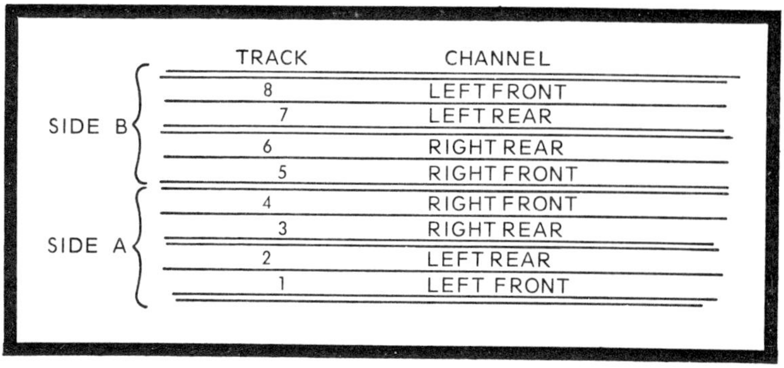

Fig. 1-8. Proposed quadraphonic track configuration maintains total compatibility. With this system, quad tapes can be played on stereo or even on mono cassette equipment. This standard has been proposed by Philips, but has met with severe industry opposition.

compatible Astrocom/Marlux system, on the other hand, maintains the same signal strength. This question has yet to be resolved, but Philips will probably prevail, since it owns all basic cassette patents.

CASSETTE TAPE EQUIPMENT

Chapter 2

The very first Philips cassette machine—the Model 150 "Carry-Corder"—which was introduced along with the cassette concept in 1964, set the standard for convenience, miniaturization, and portability. While there are many compact portable cassette machines on the market today, the most common ones are approximately the same size as the original Model 150 (see Fig. 2-1). This machine is generally a little too large to fit in the pocket, but it does go nicely into an attache′ case, and it provides a great deal of all-around convenience for business and pleasure use.

TRANSPORT

The original tape transport concept was both compact and ingenious. The drive motor was designed to draw very little battery current, previously a major stumbling block for most battery-operated tape equipment. The battery supply for the 150 and most of its successors consists of five size D cells.

The transport likewise maintains the same simplicity of design and is operated by a single-lever control for play-record, along with fast-forward and rewind functions. This same basic type of transport survives to this day in many European and Japanese recorders. The transport's best feature: its great serviceability without serious mechanical problems.

One of the major drawbacks of this type of transport from the user's viewpoint is the fact that the fast-forward and rewind functions cannot be locked into position; the control lever is spring-loaded and returns to its neutral position as soon as it is released. While this is a drawback in certain respects, it also helps to save betteries—especially if the user is absent-minded about turning the recorder off when a winding function is finished.

Slip Clutch

The basic drive train in all cassette machines includes a slip clutch. This is a device which prevents too much torque from being applied to the tape when it reaches either end. Cassette tapes have trapped leaders and trailers; the ends of the tape are permanently fastened to their respective hubs.

Many of today's home machines provide an automatic shutoff at the end of the tape. This is not a true shutoff; all it really does is turn off the motor, and in some cases flashes an amber warning light to indicate that the tape has ended. In any event, it certainly takes some of the strain off the tape and the motor. Some portable machines are also being made with this automatic shutoff feature.

Fig. 2-1. The original Carry-Corder, in this model marketed by Wollensak, embodied all of the basic design principles that were carried through for all other cassette machines to follow.

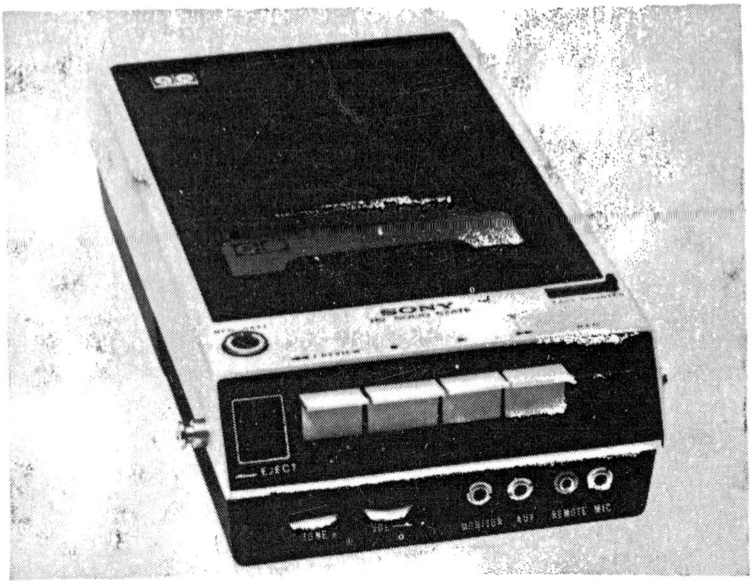

Fig. 2-2. Piano-key controls were the first major innovation in pocket-size cassette recorders, as shown in this Sony model.

Piano-Key Controls

In basic transport mechanisms, the piano-key type control is gaining in popularity. While such a control system is obviously more mechanically complicated than the single-lever knob used on the original Philips machine, the keyboard arrangement certainly makes up for this internal complexity with its convenience (see Fig. 2-2). While keyboard-controlled machines at one time brought a premium price, it is now possible to purchase such equipment very inexpensively—even for less than the single-lever controlled Philips/Norelco equipment.

However, keyboard controls do have one serious drawback; they require a certain amount of space. Because of this, virtually all of today's pocket-size recorders use a variation of the single-lever function control (see Fig. 2-3). Typically, such a pocket recorder will fit into the pocket of a large raincoat, but will make an unsightly bulge in a suit jacket. These recorders have several advantages: they nestle snugly in the palm of the hand and, with the built-in microphone, provide an ultra-compact unit for dictation while on the go. These machines also use a standard size cassette which can be transcribed by a secretary on a full-scale cassette office machine.

POCKET RECORDERS

Possibly the great granddaddy of all pocket recorders is a truly pocket-size machine made by Philips and introduced several years ago (see Fig. 2-4). This recorder does not use standard cassettes; instead, it uses a special cassette approximately one-quarter the size of the standard one, and has a recording capacity of only 15 minutes on each side of the tape.

Because there are few office machines that accept this size cassette, dictation is done with the expectation that the material will be dubbed onto a standard cassette or dictating machine through a patch cord.

Miniaturization on this scale costs money and a typical pocket-size cassette recorder retails between $70 and $100, as

Fig. 2-3. Slider-type mechanical controls are used on the 3M-Wollensak Model 400 pocket-size cassette recorder.

Fig. 2-4. The first truly pocket-size machine, the Philips "Pocket Memo" uses a nonstandard cassette, has no capstan, and runs on a single 9V transistor-radio-type battery.

opposed to $25 to $50 for a standard size cassette portable recorder. Yet, when space, weight, and portability are of paramount importance, as they very often are, the mini-cassette recorder is very often the preferred choice for businessmen.

Most of these recorders make extensive use of integrated circuits and transistors, along with some miniature capacitors and resistors mounted on a very compact printed-circuit board. Typically, the recorder's silver-dollar-size motor is a printed-circuit dc type with a fairly low speed. The motor is isolated from the drive by a thin belt which goes around the capstan flywheel (see Fig. 2-5). The belt, in most cases, looks more like a rubber band, but don't be misled by it.

Since all of this equipment is transistorized, the recorder will invaribly use a VU meter instead of the electron-ray tubes that were so popular among tube-type tape recorders a few years ago. In the case of portable machines, the meter also doubles as a battery-condition checker, with some colored portion of the meter scale indicating whether or not the battery is up to snuff.

ATUOMATIC LEVEL CONTROLS

Many of the newer recorders contain automatic level control circuits. If these circuits are used in a portable machine, very often there is no choice as to whether or not the circuit is in use. Such an instrument often will not have a VU meter. Automatic-level-control circuits are a major convenience for most types of voice recording, and once you've used a machine so equipped, you wonder how you got along without this feature.

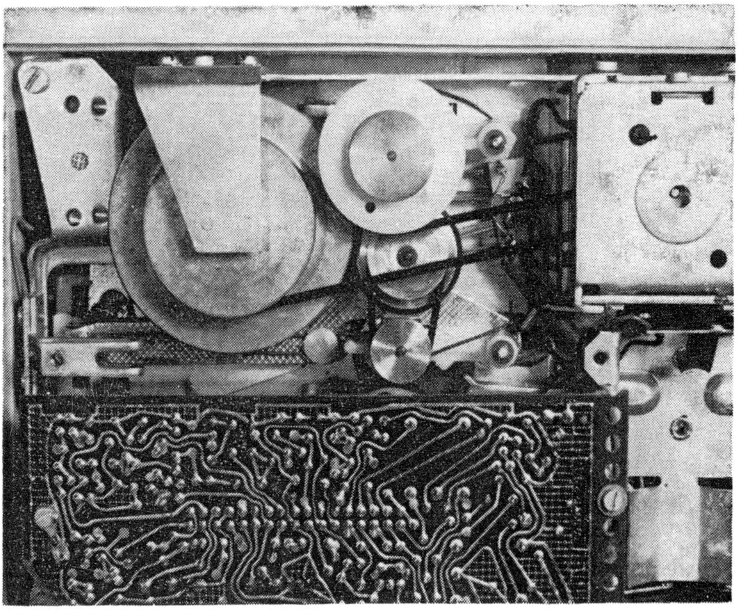

Fig. 2-5. Under the "hood" of the Carry-Corder is a simple but effective drive mechanism and ultra-compact circuit board.

While these circuits are very handy for voice recordings and the wide variations encountered in interviews and other interpersonal exchanges, the automatic level control should never be used for recording music. The reason for this is that the circuit acts as a compressor, taking loud volume passages and squeezing them down into the dynamic range of the circuitry and the tape at some preset level. This has the effect of boosting low-volume passages to high levels. In effect, a violin comes out sounding as loud as a bass drum, along with a lot of extra background and compression noise.

More expensive recorders that have automatic level control provide a method of switching it on or off. When the control is in the off position, the VU meter must be used as in conventional recordings. Some machines continue to indicate recording levels on the VU meter even when the automatic level control (alc) is turned on. This is, in effect, a "confidence indicator," since it lets you know that the recorder is indeed functioning the way it should be. Many machines either do not have a VU meter or do not provide any visual indication during recording in the alc mode. With them, you just have to keep your fingers crossed and hope everything is taping correctly. The automatic level control is covered in somewhat more detail in later chapters.

BATTERIES AND POWER SUPPLIES

Chapter 3

The very first cassette tape recorder was the Philips/Norelco 150 (see Fig. 3-1); in fact, it set the pattern for everything that followed. It not only set standards for the cassette format, but is also was the first true battery-operated portable. There had been portable tape recorders before, but they required massive batteries and used them up in a hurry. The first cassette recorders pointed the way to total portability and low battery drain.

BATTERY CONSIDERATIONS

In selecting a battery, there are several considerations: size, weight, ampere-hour capacity, and durability in various applications. The first Philips portable recorder used a motor that was designed to have relatively low battery drain. At the same time, the battery had to be designed to provide at least enough operating time to keep battery cost down and the inconvenience of changing batteries to a minimum. The Norelco 150, along with many of its successors made by manufacturers all over the world, uses five size C cells. As of this writing, this represents a retail price of between 75 cents to $1.50—relatively inexpensive considering the amount of operating life that this set of batteries provides.

Some larger recorders, such as the Wollensak shown in Fig. 3-2, use size D cells, which are much larger and heavier. The idea here is that the machine is not all that portable to begin with; and if it is to be used as a portable, it should have a long-lasting battery supply. Unlike the Norelco 150, this particular Wollensak recorder contains a built-in ac converter so that it can be used easily as a portable or an ac-powered machine with only the flip of the switch.

Battery Size vs Life

Some of the newer pocket-size portables such as the Wollensak 400 shown in Fig. 3-3, use four penlight (AA) cells—

Fig. 3-1. Battery supply for the first cassette recorder—the Carry-Corder—was and still is a cluster of five size C cells.

substantially reducing both the size and the weight of the tape recorder. It also reduces the length of service possible from a set of batteries; therefore, frequent battery changes are necessary with this type of machine. Here, battery economy is sacrificed in favor of extreme portability.

AC Supplies

Almost all of today's batttery portables have some kind of provision for using ac current. In some machines, the ac conversion or power supply circuitry is built into the equipment. In the smaller and more portable machines, such circuits must be external, and usually take the form of an ac adapter which plugs into a wall outlet and into the tape recorder (Fig. 3-4). Some tape recorders may not have this provision, usually the older models.

By using the ac adapter (which is either optional or provided as part of the kit that comes with the recorder) it's possible to extend battery life considerably—especially for machines that use penlight cells. Another type of power supply uses power from an automobile battery by plugging into the cigarette lighter. This type of adapter is shown in Fig. 3-5.

BATTERY TYPES

When the time comes to buy batteries for your tape recorder, the perennial question is, which kind? There are three general types of battery to consider; the fourth type is mentioned here because it should be avoided. The three classic groupings are (1) carbon-zinc; (2) manganese-alkaline; (3) nickel-cadmium rechargeable. A fourth kind of battery is the mercury type; however, for reasons that we go into later, this type should not be used in tape equipment.

Carbon-Zinc Battery

The basic carbon-zinc battery is probably the most familiar type and is the kind of battery most commonly used

Fig. 3-2. Larger units take larger batteries, such as this Wollensak with its D cells. It also operates on ac.

Fig. 3-3. The new breed of pocket portables, including this Wollensak 400, take lightweight AA (penlight) batteries.

today. Typically, a size D carbon-zinc cell retails for 25 to 30 cents and will provide a fair amount of operating time under most normal load conditions. This battery is a well proven as well as a much-maligned type.

Many battery companies seal their carbon-zinc cells in steel and claim these batteries are leak-resistant. Certainly, such cells are less likely to damage flashlights and tape recorders than unsealed ones. But remember, like all batteries, they contain a continuing chemical action inside that steel jacket; it left long enough, the steel case will rupture and leak all over your valuable equipment.

The moral of the story is: change carbon-zinc batteries frequently and keep a record of the last changing date as shown in Fig. 3-6. A self-stick label somewhere inside the recorder's carrying case or even on the recorder itself will serve as a handy reminder of just how old those batteries are that you thought you put in only last week or so.

The carbon-zinc battery consists of an outer shell made of a zinc can which is also the negative electrode (see Fig. 3-7). The positive electrode is a stick of carbon in the center of the can and separated from it by the electrolyte mix. The electrolyte is a paste made of ammonium chloride and zinc chloride dissolved in water. The positive electrode is actually manganese dioxide, a chemical coating which forms on the carbon center pole. The carbon itself acts as a catalyst and a conductor for the positive pole.

Fig. 3-4. Easy way to conserve batteries is to use an ac adapter such as this one supplied with the Wollensak 400.

Fig. 3-5. An alternative power source for the Wollensak 400 is this cigarette-lighter adapter for highway use.

The carbon zinc battery is correctly known as a Leclanche cell. Typical of this type of cell is the large, old-fashioned dry battery of the type we used to use for electrical experiments in grammar school. In some rather ancient homes, electric door bells are still operated by huge dry cells of this type, although it's a rather expensive way to ring a chime in the front hall. Variations on the basic carbon-zinc cell include batteries specifically designed for flashbulbs, transistor radios, and other special applications. All that's different are small changes in the way the electrolyte paste is mixed. Of course, some cells of this type are better for certain applications than others, but the casic carbon-zinc cell is pretty much interchangeable.

Alkaline Battteries

The alkaline battery is much more of a newcomer on the scene. It is actually a manganese-alkaline type, which is very actively promoted these days because of its extra-long operating life. You may see fantastic advertising claims to the effect that this battery will provide up to ten times as much life as the regular "old-fashioned" type of battery, but the most an alkaline cell can provide is perhaps two to three times as much operating life as a carbon-zinc type.

For this extra operating life the manufacturer charges anywhere from three to four times as much for the battery. A typical D cell, instead of costing 25 cents, costs 89 cents or more.

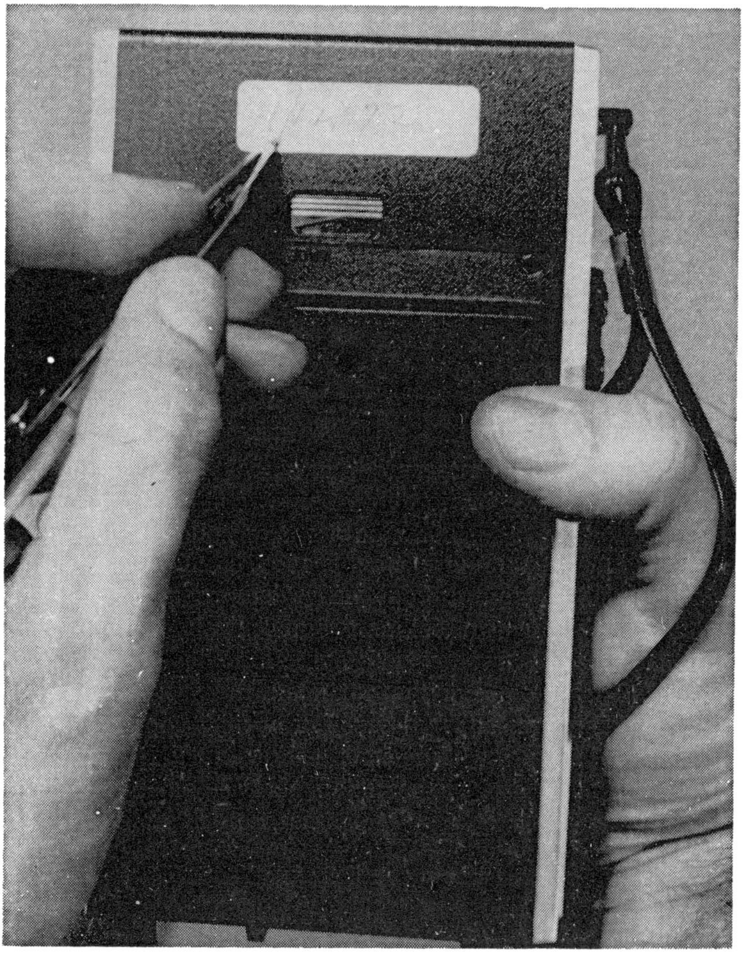

Fig. 3-6. Be sure to enter the date fresh batteries are installed on a label on the recorder; otherwise you may get caught with your batteries down.

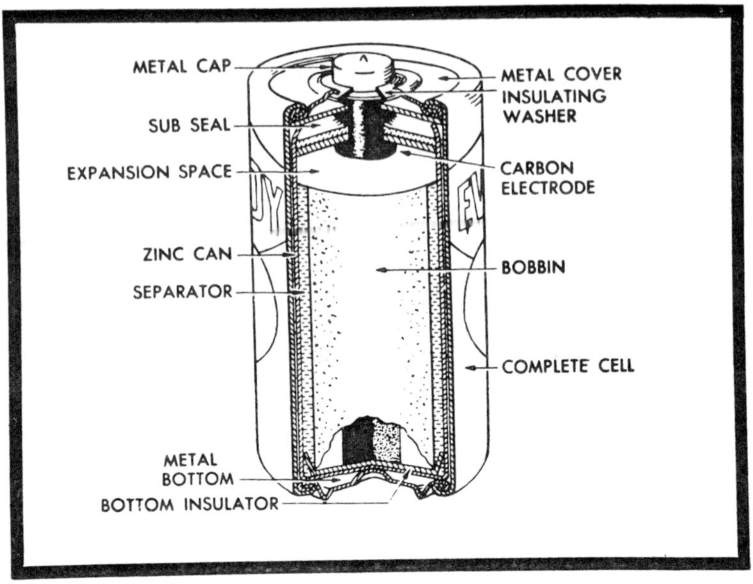

Fig. 3-7. Cross section of common carbon-zinc (Leclanche) cell in the popular size D configuration. This type of battery continues to be the most popular for general use.

One of the problems with alkaline cells is that they tend to leak in a very inconspicious way. A thin film of leakage oozes out of the case and covers the negative electrode area with a thin film of white powder. While the battery is still fine, this white powder can cause corrosion damage to the equipment; it causes an even more insidious problem—it ruins the electrical contact between that point and the positive cap of the next cell or of the power terminal in the battery compartment of the recorder. The only cure for this is to sandpaper the affected tape recorder and battery terminals. With problems like this, it is very often a good idea to stay away from the alkaline type, especially if it's going to sit in the tape recorder for a long period of time.

The alkaline doesn't look different from the zinc-carbon, at least on the outside. Its negative terminal is connected to a series of zinc plates internally (see Fig. 3-8). Once again, zinc is the negative electrode, although in this case it must be connected separately to the outer shell, which actually forms the contact with the bottom of the battery. The positive terminal is a large area of manganese dioxide and the electrolyte is calcium hydroxide. The battery is physically and electrically interchangeable with the carbon-zinc battery, since its output voltage characteristics are identical and it is available in exactly the same sizes.

Somewhat less well known is a rechargeable version of the alkaline battery. You can expect to pay more for this battery, and its rechargeability is limited to only a few hundred charge-discharge cycles at the most. The rechargeable feature is possibly useful for some applications, since the battery's initial cost is lower than the more commonly used nickel-cadmium rechargeable battery, which is described later.

Mercury Battery

The fourth type of primary battery is the mercury cell, but it is not electrically interchangeable with carbon-zinc and alkaline types because its initial output voltage is lower—about 1.2V as compared with 1.5 for the other two types. For the mercury battery to be used effectively, equipment must be designed for this lower voltage, and this certainly reduces your options for using other battery types. Mercury batteries are also quite expensive as compared with alkaline and carbon-zinc types. What then is the reason for its being?

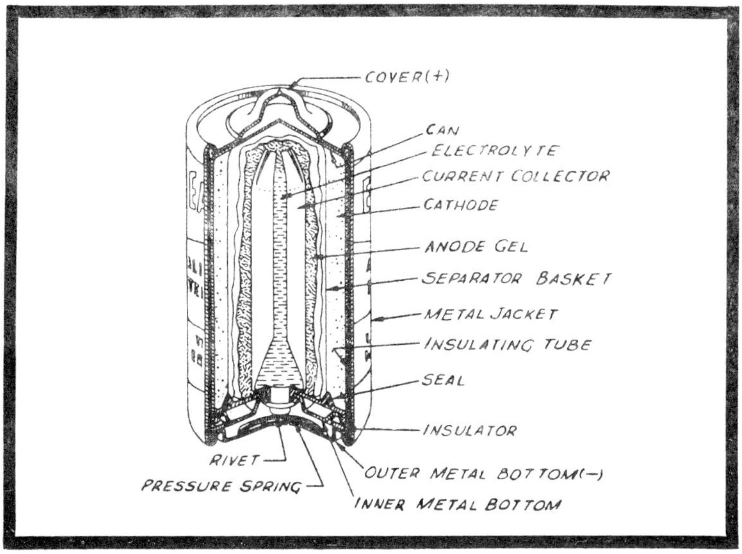

Fig. 3-8. Cross section of a manganese-alkaline size D cell—a battery type that is growing in popularity, but special problems are associated with its use.

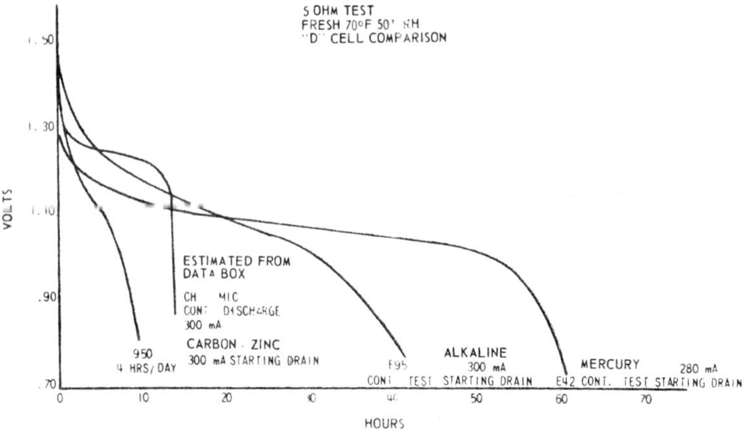

Fig. 3-9. Discharge curves into a 5-ohm load for D size cells show comparative lifespans for carbon-zinc, alkaline, and mercury batteries. (Courtesy Union Carbide.)

For one thing, it has a much longer life span than either the zinc-carbon or the alkaline. Unlike the other two batteries, which slowly lose their voltage as they deteriorate and are used up, the mercury battery retains the same operating voltage until it is nearly exhausted, and then drops off quite suddenly. This may be an advantage in some recorders, and certainly a disadvantage in others. It's all a question of what you need.

The discharge curves of typical size D batteries are shown in Fig. 3-9. Notice the long and almost level slope of the mercury cell's discharge curve and then the knee in the curve as it drops off rapidly.

Secondary Batteries

The batteries we've been talking about here are called **primary** batteries because they are generally not rechargeable; once they have been exhausted, they can't be reused. Any battery that can be recharged and reused is called a secondary battery. Probably the best known secondary type is the one in your car—a lead-acid battery which requires periodic and careful attention to keep it operating. It's also a bit cumbersome and sometimes even dangerous, since it uses sulphuric acid—something you may have discovered the hard way.

The nickel-cadmium battery, on the other hand, is modern science's contribution to the secondary battery family, and it is a remarkable device indeed. It is rechargeable through thousands of charge-discharge cycles, is sealed nicely into a leakproof case, never needs to have its electrolyte level checked or changed and never needs water added. It is directly interchangeable with carbon-zinc and alkaline batteries, since it provides the same nominal 1.5V per cell that they do, and is manufactured in many of the same sizes and shapes.

With such a great battery on the market, you may wonder why anyone bothers with anything else. N-C batteries do have their problems. For one thing, they tend to leak off and lose their charge after a few weeks of nonuse. This means that if your tape recorder has nickel-cadmium batteries in it and it's been sitting in the drawer for a month, when you pull it out and want to use it again it won't work. Generally speaking, it's a good idea to plug in a charging unit and give these batteries a "goose" every week or so to keep them fresh.

Other than that, the nickel-cadmium is a standout performer and really does an admirable job. It is so reliable that it is commonly sealed inside expensive electrical devices such as electronic flashes and electric razors. Its power potential is so great that today we even have electric drills and other shop tools that are powered by these batteries—all sealed in! A typical size D nickel-cadmium cell is shown in Fig. 3-10.

Another very special feature of the nickel-cadmium battery: it's almost impossible to overcharge—very much unlike its automobile lead-acid counterpart. This is because of the mix of materials and chemicals within the battery, which absorb gas (oxygen) produced by the charging cycle. The gas simply combines chemically with the other materials in the battery to form an inert compound. This feature is especially good for the absentminded hobbyist who leaves the charger plugged in too long. It's also very handy for anyone who uses rechargeable batteries and can't sit with a timer or a clock to watch for the precise charge period. An overnight charge is sufficient in most cases.

RECHARGING PRIMARY BATTERIES

There is a school of thought concerned with recharging primary batteries. Actually, it isn't really a case of recharging

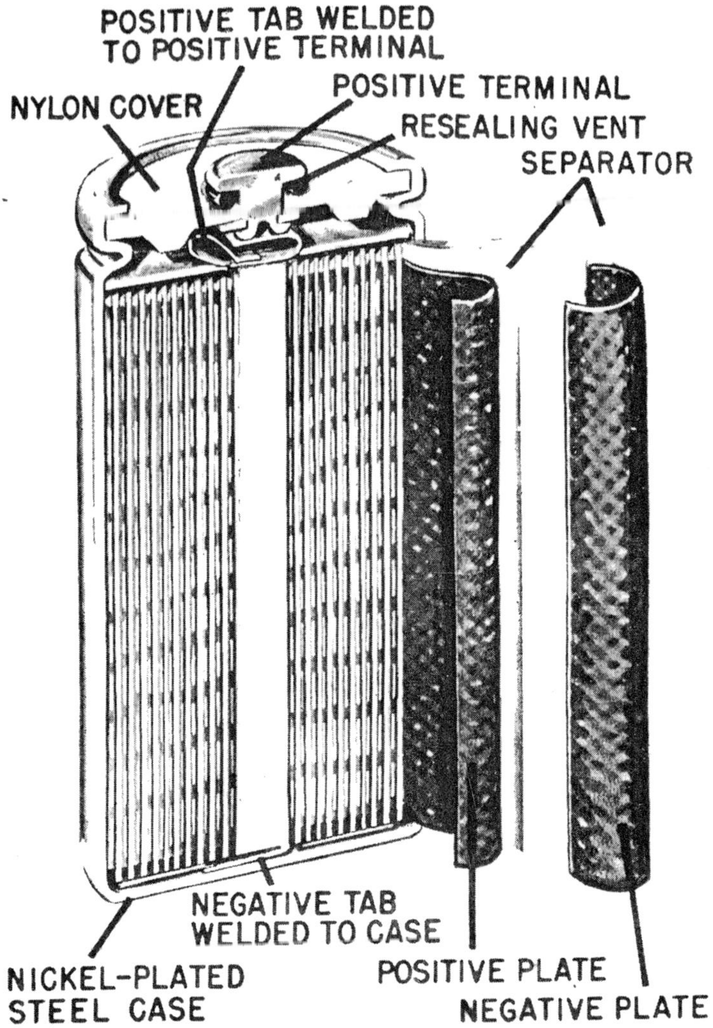

Fig. 3-10. Nickel-cadmium D cell cross section. (Courtesy General Electric.)

the battery and putting back the chemicals in the original condition. It just doesn't work the way it does with secondary batteries. Rather, it is a reversing of **some** of the chemical deterioration that has taken place in the battery by a controlled amount of reverse current flow.

Yes, it's possible to extend the life of primary batteries with special charging circuits, but it's also a very dangerous thing to do because a couple of minutes too long on the charger and these batteries can explode, with horrendous results. A typical battery charger that can be used for both charging nickel-cadmiums and for rejuvenating primary cells is shown in Fig. 3-11.

Fig. 3-11. Common home-type battery charger can be used for recharging nickel-cadmium types and for extending the lifespan of primary batteries.

SHELF LIFE

An important point to consider when buying primary batteries is shelf life. All batteries are electrochemical devices, and the chemical reaction continues in the battery even though it's not being used. An imported battery may have been sitting on a Japanese ship for six months or on a dock in San Francisco before being shipped to the east coat where it is eventually sold.

With this amount of time spent in storage and shipping, the battery has already lost a great deal of its useful life, as much as a year or more, before you purchase it. We're not saying don't buy imported batteries; just be careful when you do, and buy them from a reputable dealer whom you feel has not had them sitting on his shelf for a long time. Imported batteries are far less expensive than domestic ones, but a 50 percent saving in price isn't worth a 75 percent reduction in operating life, which could well be the case.

On the other side of the coin, it's quite possible to buy American-made batteries that have been sitting on the shelf a lot longer than their Japanese counterparts, and that sometimes may be completely dead at the time of sale. Just use discretion, know your dealer, and figure your prices carefully when buying.

TAPE FORMULATIONS AND RECORDING CIRCUITS

Chapter 4

In the quest for better recorded sound, manufacturers have for years tried many variations on the basic ferric oxide that is the cornerstone of all tape coatings. At about the same time that cassettes first appeared, 3M introduced its Scotch brand low-noise tape, which soon came to be known as "Dynarange."

This particular formulation had more closely spaced oxide particles and at the time it was claimed that this tape would provide excellect frequency response, even at the slow open-reel tape speed of 3¾ ips. This meant that twice as much material could be recorded on the same length of tape that would normally be recorded at 7½ ips.

The importance of this tape was not lost on the cassette industry. Low-noise formulations from 3M and other manufacturers soon proliferated, and this tape and similar types have been used almost universally in all but the cheapest of cassettes.

But even using this low-noise tape isn't enough. The ultraslow speed of the cassette (1⅞ ips) naturally affects frequency response and noise level considerably. Typical frequency range of early cassette machines was about 100 to 8000 Hz. Some machines were able to coax as much as 10 kHz out of the cassette, but they had to strain to do it.

NEW TAPE OXIDES

Newer arrivals on the tape scene in the last couple of years include doped oxides and gamma ferric oxides from various manufacturers, and these provide improved frequency response. During the same interval, DuPont started licensing manufacturers to produce and market its Crolyn tape—a chromium dioxide coating that can produce vastly superior frequency response in cassettes. But some of these new tapes require special circuitry in the machines. Very often, people don't understand how to use them and try to make them work with the wrong kind of equipment.

The most extreme case came with the introduction of Cobaloy by Graham Magnetics. The coating is not an oxide at all, but a cobalt alloy, and it requires vastly different recording electronics in any machine that uses it. At this writing, this latest type of tape is not yet available in audio cassettes.

RECORDING BIAS

Today's ferric oxides, particularly the doped and gamma types, are able to produce extraordinarily good recordings, but again the bias electronics must be just right.

Bias is a high-frequency signal, roughly between 80 and 105 kHz, mixed with the incoming audio signal when a recording is being made. This is done because magnetic tape has one serious drawback: it's nonlinear at treble frequencies. If you try to record a straight audio signal without bias, you lose all the high frequencies. The bias signal, mixed with the incoming audio, has the effect of getting rid of this nonlinearity.

One way of mixing bias, employed by some equipment manufactuers, is the use of cross-field heads. Instead of mixing the bias electrically with the incoming signal, the bias is fed to a special head which faces the recording head from behind the tape (see Fig. 4-1). The bias field from this head mixes with the recording signal field from the record head, and the mixing takes place in the tape itself.

Bias is handled differently by different tape recorder manufacturers. The choice of frequency depends on the erase oscillator frequency, since the erase oscillator also supplies the bias signal (see Fig. 4-2). Whatever frequency is used for erasing becomes the bias frequency as well, and this is often governed simply by the economics of recorder design.

Bias Drive

In addition to the bias-frequency consideration, there is bias drive which is the amount of voltage or signal strength of the applied bias. This depends on the type of tape that is used. Some tapes require more bias drive than others.

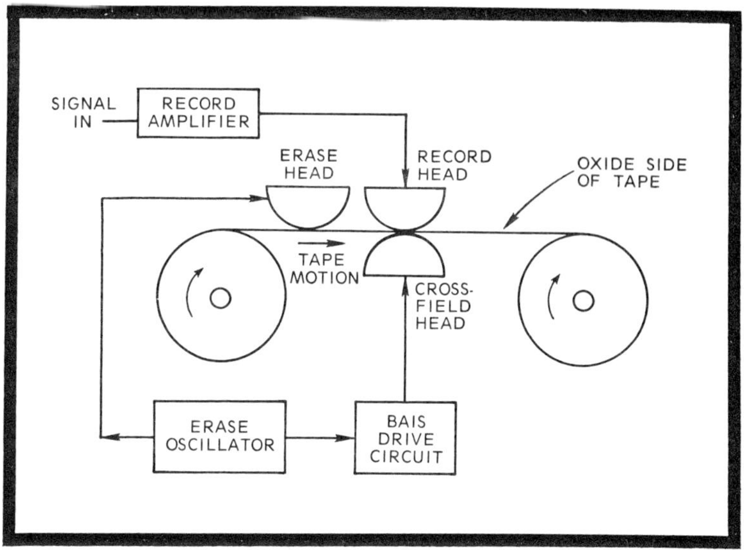

Fig. 4-1. One method of mixing the bias and audio signals is to feed bias to a separate head behind the tape. The bias head's field mixes with the record head's magnetic field within the tape itself. This setup is called "cross-field" heads.

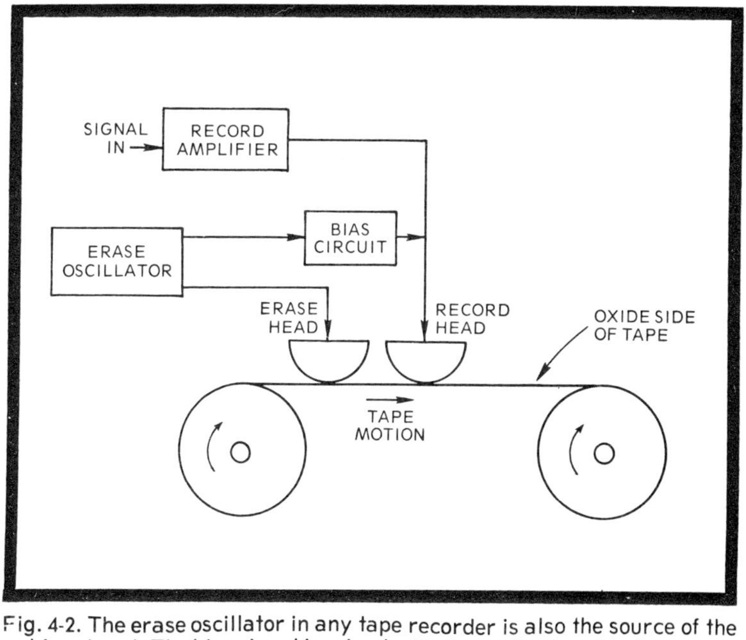

Fig. 4-2. The erase oscillator in any tape recorder is also the source of the ac bias signal. The bias signal is mixed with the audio signal in the record head either through separate windings or feeding the same winding.

The bias signal is mixed with the incoming audio signal right at the record head. Since it's usually just a short wire length from the erase head, some small circuit components are interposed to adjust the actual bias drive level, which must be a small faction of the erase signal. It is at this point, and with this adjustment, that the recorder is finally set at the factory to mate to a particular kind of tape coating. Most recorders today are set to operate best with some sort of low-noise tape.

Bias is just as important for cassettes as it is for open-reel tapes. With any quality cassette deck, some tapes provide better results than others. The ones that work better presumably are closest to the tape type that the machine bias was set for at the factory. But what tape did the manufacturer use? Few if any recorder manufacturers reveal this information. About the only concrete clue is to note carefully the kind of tape that was factory-packed with the machine. This is usually the kind of tape that the bias has been set for. But even this clue may not be totally accurate. Some manufacturers will take a perfectly good tape and affix their own label on the cassette. Since the company makes recorders and not tapes, you can search through all the stores in your area and not find another cassette with this label, except factory-packed with another like-model machine.

Another major problem is the demonstrator or shelf display model, which is the last machine in stock. The factory-packed tape may have completely disappeared, so the dealer gratuitously supplies a blank cassette from stock, without so much as a second thought. There goes your clue! When in doubt, ask. You may be rewarded with a blank stare. Or the dealer will say, "Oh, any brand of tape will work with this machine." Sure they will, but how well?

Bias Selector Switch

Some cassette machines have a bias selector switch. Bearing such cryptic labels as "Iron Oxide" and "Chromium Dioxide" or "Fe_2O_2" and "CrO_2" (see Fig. 4-3). This switch enables the operator to record on the new chromium dioxide tapes with the appropriate bias applied. This is a situation that is somewhat unusual, since chromium dioxide requires a bias drive that is so vastly different that the industry has become acutely aware of this need.

For hobbyists who don't happen to have such equipment, or who do and still prefer ferric oxide, it's quite possible to obtain excellent results with stereo cassette recordings. For one thing, no matter what you may think, **never** use that bias switch in the "CrO_2" position for ferric oxide tapes. Ferric oxide and chromium dioxide are so widely different in their biasing requirements that you'll just end up with garbage on the tape.

Some cassette machines are now on the market with adjustable bias, and in a couple of cases have a bias selector switch for "regular" and "premium" ferric oxide tapes. However, such machines are still the exception, and the question remains, "How well will premium tapes work with my existing tape deck?" Depending on which tape you're talking about, the answer can be "fair," or it could be "quite well."

If the machine is biased for a "standard low-noise" tape, as most decks are today, chances are the bias setting will be somewhere within the range of possible optimum bias settings recommended for premium tapes. In checking such tapes with

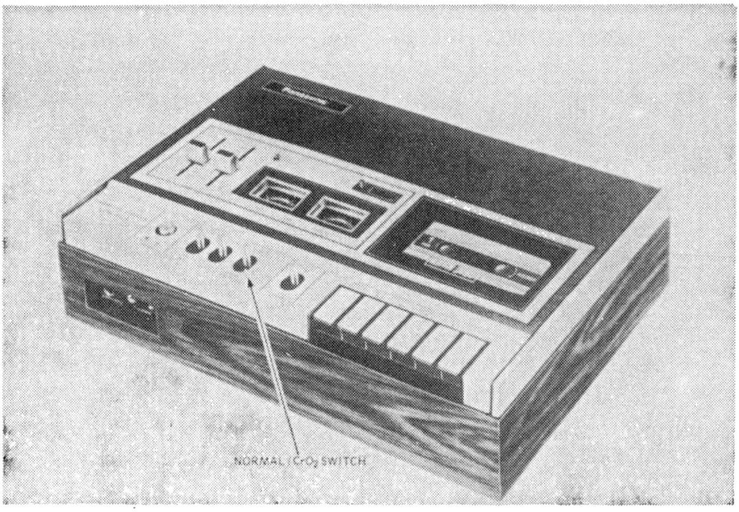

Fig. 4-3. Some machines are equipped with a hi-bias/lo-bias switch. In most cases, these switches pick between ferric oxide and chromium dioxide bias settings, as is the case in this Panasonic unit.

a variety of home cassette decks, it was found that most of them record extraordinarily well with most quality decks. Further, in an effort at industrywide standardization, decks coming out of Japan now are biased for the identical premium tape (TDK's SD) and for other tape formulations that have similar bias requirements.

Bias and Equalization

Let's assume that you have available to you one of those rare, new machines that allow you to select "standard" and "premium" tape bias for recording. Setting that bias switch in the "premium" position may not just affect the bias. There are several schools of thought on the best way of squeezing more fidelity out of cassettes, so that switch may also control equalization.

The three basic approaches to this are: (1) optimize bias only for the particular tape; (2) optimize bias and recording equalization; and (3) optimize bias and playback equalization. If you happen to record on a machine with condition 2 and play back on a machine with condition 3, there may be a problem, and you'll end up diddling with the treble controls on your amplifier, hoping that you can undo the damage. For that matter, even if your cassette equipment doesn't have a bias or equalization control on it, you can still do some tone control juggling on your amplifier. Some preamps even have graphic equalizer controls for reshaping the signal. That would be the idealized way of correcting for recording equalization.

DOLBY NOISE REDUCTION

An important part of quality home stereo equipment today is the Dolby noise-reduction system. Dolby has been used for several years now by the recording industry for making masters. It's only in the last couple of years that it has been recognized as an important part of home music systems as well.

The Dolby system can reduce tape noise (hiss) by as much as 10 dB in cassettes. What this means in terms of home equipment is a vast improvement in signal-to-noise ratio. Background noise is very distressing, especially during low passages of music. In the past, we have had to live with high hiss levels coming from the tape; it's a natural characteristic of the tape and the equipment that we use with it. What the Dolby system does is to expand the musical material during quiet passages (when hiss is most noticeable) while leaving very loud music passages relatively untouched. In playback, just the reverse happens; the low-level expanded music passages are compressed back to their original format. At the same time, noise level is pushed down several dB by this expansion and compression so that it disappears below the level of audibility during most of the low-level music passages (see Fig. 4-4). Suddenly, the recording is totally hiss-free—an unbelievable experience when compared with conventional tape listening.

Open-reel equipment for home use typically has a signal-to-noise ratio of 50 to 55 dB. In exceptional cases, it may go as high as 60 dB, which is really comparable to professional studio equipment quality. The best home cassette equipment typically has a signal-to-noise ratio of 40 to 45 dB. By providing an extra 8 to 10 dB of improvement in this figure, we suddenly find cassette equipment with a signal-to-noise ratio comparable to that of the best open-reel home units. This is what Dolby is all about, and this is why duplicating companies are now Dolbyizing tapes on their duplicating lines for home use. A diagram of a typical Dolby cassette unit is shown in Fig. 4-5.

When playing back through the appropriate circuitry, Dolby tapes provide startling results. But if your equipment

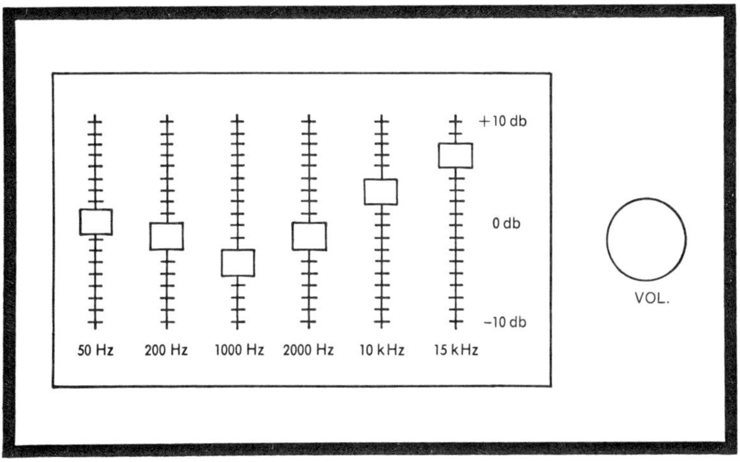

Fig. 4-4. Slider controls of a graphic equalizer show at a glance just what type of playback curve is being used for the recording.

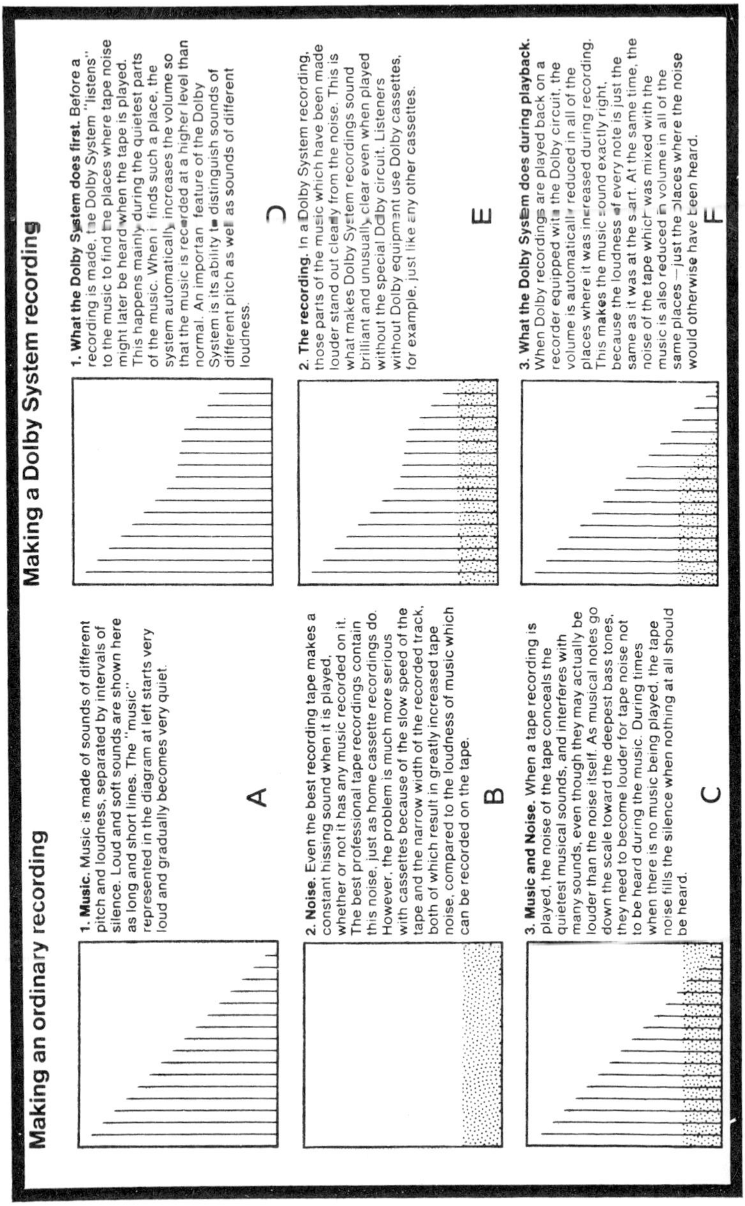

Fig. 4-5. The signal level of an ordinary recording (a) is superimposed on hiss noise (b) with the result that low signal levels disappear into this "hash" (c). When Dolby B processing circuitry is used, the signal level is altered (d) so that low levels are higher than the noise level (e). When played back with normalized level (f), the hiss is reduced below the lowest low-level signal.

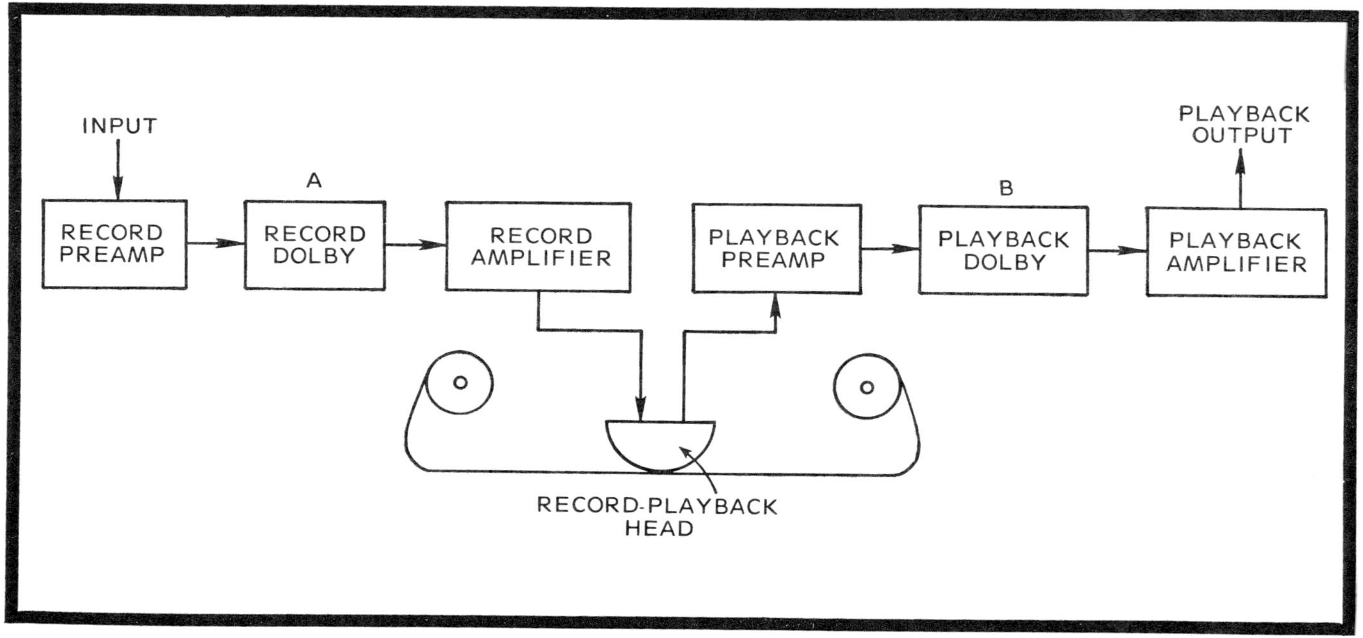

Fig. 4-6. How the Dolby noise-reduction circuit is used in a cassette recorder. The same Dolby circuit board (blocks A and B) is used via switching for both recording and playback. A cutout switch bypasses the Dolby entirely when desired.

Fig. 4-7. An add-on Dolby, the TEAC AN-50, is moderately priced and can be used for all tape systems, cassettes, cartridges, and open reel. By using this building block, existing quality cassette decks can be updated without being replaced.

doesn't have Dolby and you're faced with playing back Dolbyized cassettes, don't be too concerned. Simply back off the treble control a little bit and your audience won't be any the wiser.

Generally, however, a Dolbyized tape played on non-Dolbyized equipment will tend to produce more hiss than a conventional, non-Dolbyized tape would. This again is something that you'll have to live with if you want the benefits of low-noise tape on some equipment and yet use it with all other types of playback units. If the hiss level gets too bad, you can back off the treble control as far as it will go (and lose some of your high frequencies along with the hiss).

There are several add-on Dolby units on the market, such as the type shown in Figs. 4-6 and 4-7. With this kind of equipment, it's possible to have the convenience of Dolby noise reduction with any tape system component. Hooked up as shown in Fig. 4-8, an add-on system like this will update almost any cassette stereo deck. The only thing it can't do is change the bias setting.

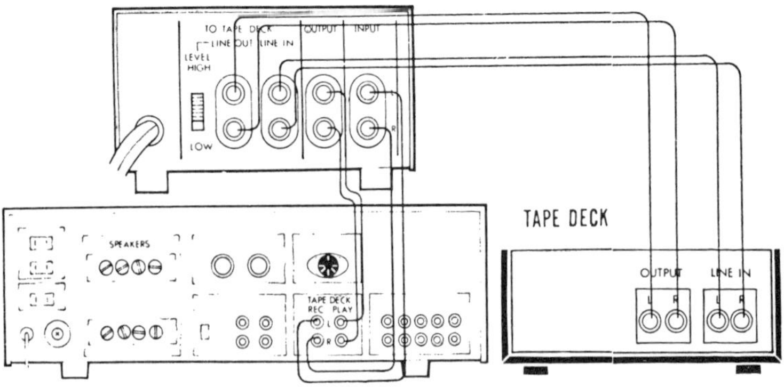

Fig. 4-8. Method of connecting a TEAC AN-50 Dolby "black box" into the circuit for tape recording and playback.

HOME STEREO EQUIPMENT

Chapter 5

Extremely important to the rapidly growing development of tape cassettes is the use of the cassette as the heart of home stereo recording and playback systems. When the 4-track stereo cassette format was first introduced several years ago, the cassette was a long way from being anything vaguely resembling a high-fidelity recording medium. Frequency response was sadly lacking—typically usable to only 8 kHz or 10 kHz; wow and flutter specifications were incredibly bad; and signal-to-noise ratio was typically 40 to 45 dB—leaving a great deal to be desired.

In the short span of two years, the picture changed drastically. New tape formulations, new circuitry for noise reduction, and vastly improved equipment and cassette mechanical performance characteristics placed the once lowly cassette in the front rank of home stereo performers.

To be sure, the cassette is still a long way from rivaling open-reel tape running at 7½ ips (inches per second). But quality has improved so much that a modern tape cassette dubbed from a good open-reel tape on a home system can sound every bit as good to the unpracticed ear as its higher fidelity counterpart. In fact, the only way to spot the difference for certain is to run both the original and the cassette copy at the same time and switch from one to the other. This type of A-B testing is commonly used in other areas of high-fidelity comparison studies, and is just as valid here. The loss of certain key high frequencies becomes apparent, but the amazing fact that so much is coming out of this tiny cassette underscores the incredible improvements that have been made.

DECK vs COMPLETE SYSTEM

When it comes to selecting equipment for the home cassette stereo system, the classic dispute over a complete

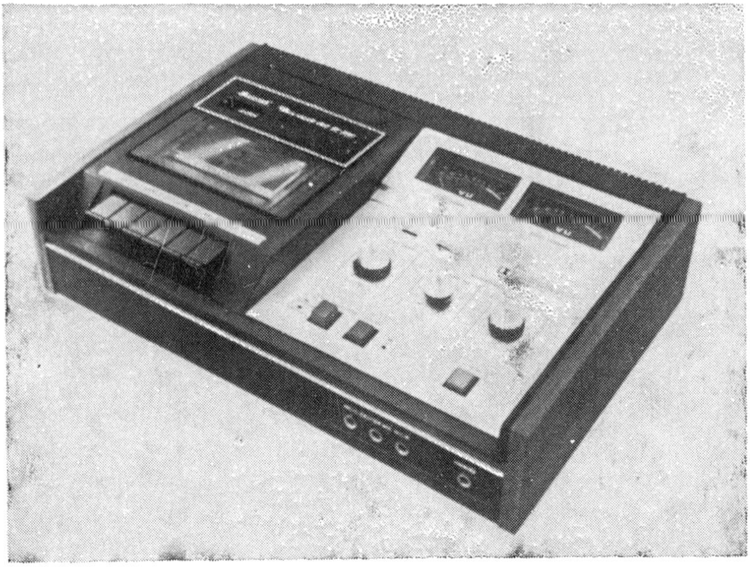

Fig. 5-1. Typical of the new breed of sophisticated cassette decks is the Sansui SC-700. Special features include switchable Dolby, switchable bias, center mic input with separate level control, input mixing, and automatic shutoff that disengages the capstan pinch roller and heads.

system versus a deck comes into consideration again. The deck is certainly far more versatile than a complete recorder-amplifier-speaker package and has the advantage of being used as a component in large-scale stereo systems.

A typical cassette deck with switchable bias and built-in Dolby noise reduction circuits is shown in Fig. 5-1. A deck of this type is not cheap. It retails in the neighborhood of $220 to $275, but produces an incredible amount of sound quality when compared with the significantly lower cost deck ($90 to $150) shown in Fig. 5-2.

Switchable Recording Bias

The problem in selecting such a high-quality deck is that there have been no hard and fast standards set for the industry on bias and equalization for new tape formulations. As indicated in the previous chapter, the variety of these new tapes has proliferated so that we no longer have a clear-cut choice between gamma ferric oxide and chromium dioxide. The earliest machinc to feature switchable bias simply offered a choice between these two tape coatings.

But with the introduction of cobalt-modified formulations for higher energy output, along with extended frequency response, and the evolution of various schools of thought on biasing and equalization, there is simply no either-or choice available today. Add to this other new tape coatings that are on the horizon, such as Graham's "Cobaloy," and the prospective cassette purchaser is left in a quandary.

Possibly the most prudent course is to purchase a machine of good quality that is accessible for later modifications. A good choice in a case like this would be something like the Heathkit cassette deck (see Fig. 5-3) which, because it is in kit form, offers the hobbyist the opportunity to open it up and make later modifications.

Built-in Dolby

Another point to consider is the fact that purchasing a machine with built-in Dolby circuitry locks the user into this type of machine which may not have the proper type of biasing for next year's new wonder tape. Again, if a Dolby is deemed necessary, and it does provide a very significant reduction in tape noise, then purchasing a separate Dolby unit of the type shown in Fig. 5-4 would possibly solve the problem. This way the extra money spent for Dolby circuitry does not go out the

Fig. 5-2. The Sony Model 127 offers switchable bias and a noise limiter (not Dolby) among its special features. Several companies have incorporated their own noise limiters into their equipment; unfortunately, they're not all compatible with one another.

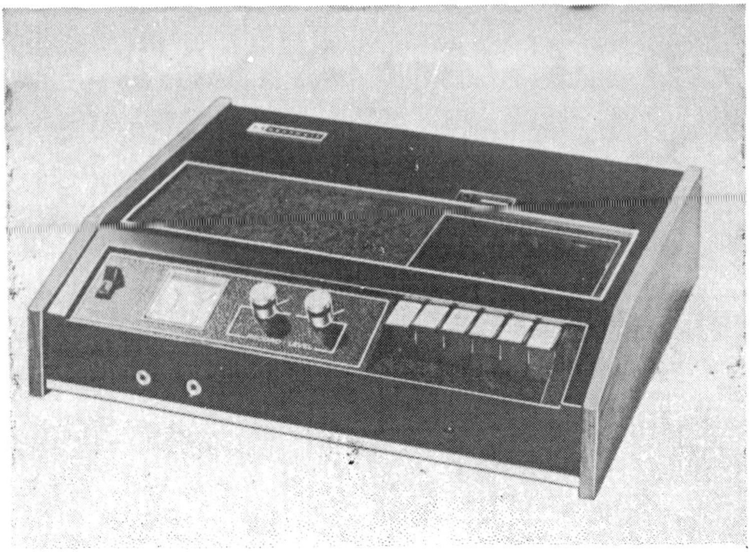

Fig. 5-3. The first stereo cassette deck to be available in kit form—the Heathkit AD-110—is a basic unit, without switchable bias or Dolby. Later Heathkits include Dolby and bias switching.

window when a machine becomes obsoleted by new formulations. A typical setup using such a system is shown in Fig. 5-5. By treating the Dolby noise reduction system as a component rather than part of the cassette deck, it can be readily removed and used with other equipment, including open-reel recorders.

How significant is the Dolby? When noise figures are compared with those available on open-reel tape, it certainly looks very impressive. A typical cassette deck without a Dolby has a signal-to-noise ratio of approximately 40 to 45 dB. A typical open-reel deck operating at 7½ ips has a signal-to-noise ratio of about 55 dB—about 10 dB better than a cassette machine. This noise level is represented by hiss.

A Dolby B unit can improve the signal-to-noise ratio anywhere from 8 to 14 dB. Thus, the Dolby unit improves cassette noise levels to the point where they are comparable to or slightly better than the signal-to-noise figures for open-reel equipment. But there is still some hiss present, just as there is with open-reel tape. However, such a Dolby unit has built-in versatility; it can be used to improve the noise figure for an open-reel tape, pushing it to the realm of 65 to 70 dB! At this kind of noise level, hiss is virtually nonexistent.

Dolby Operation

The Dolby system itself is possibly one of the greatest boons to the recording industry since the invention of magnetic tape. Noise added by a tape recorder to a total recording has always been a problem. The Dolby system basically expands the low or quiet passages of music—where hiss or other tape noise is most obvious—to a relatively high volume level during recording. By the same token, loud passages of music, which tend to mask any extraneous noise, are compressed in volume level during the recording.

During playback the process is reversed; the compressed loud passages are expanded to their original level while the quiet music passages are compressed to their original level. In the process of compressing or pushing down the volume level of the quiet passages, any recorded noise present on the tape is pushed down by a like amount, but in this case it is usually below the threshold of human hearing. This is shown graphically in Fig. 5-6.

Recognizing the importance of this development and the need to decrease the recorded noise on tapes, the recording industry has been using Dolby processors for several years now in the manufacture of its master tapes. The result has been remarkably quiet recordings, and the final disc made from a Dolbyized master is virtually noise-free. Prerecorded tapes made from such masters naturally must be non-Dolbyized. Of course, they pick up a certain amount of hiss during the dubbing process, but is nowhere near the amount of hiss and other noise that would have been present without the Dolby processors in the mastering.

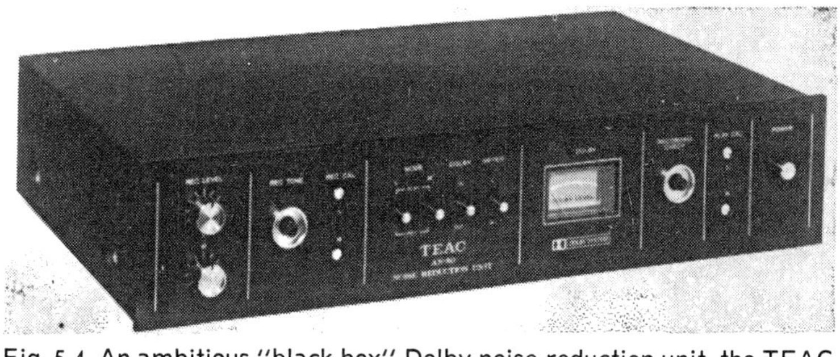

Fig. 5-4. An ambitious "black box" Dolby noise-reduction unit, the TEAC AN-80 is designed for sophisticated home systems and can be used with any type of recorder.

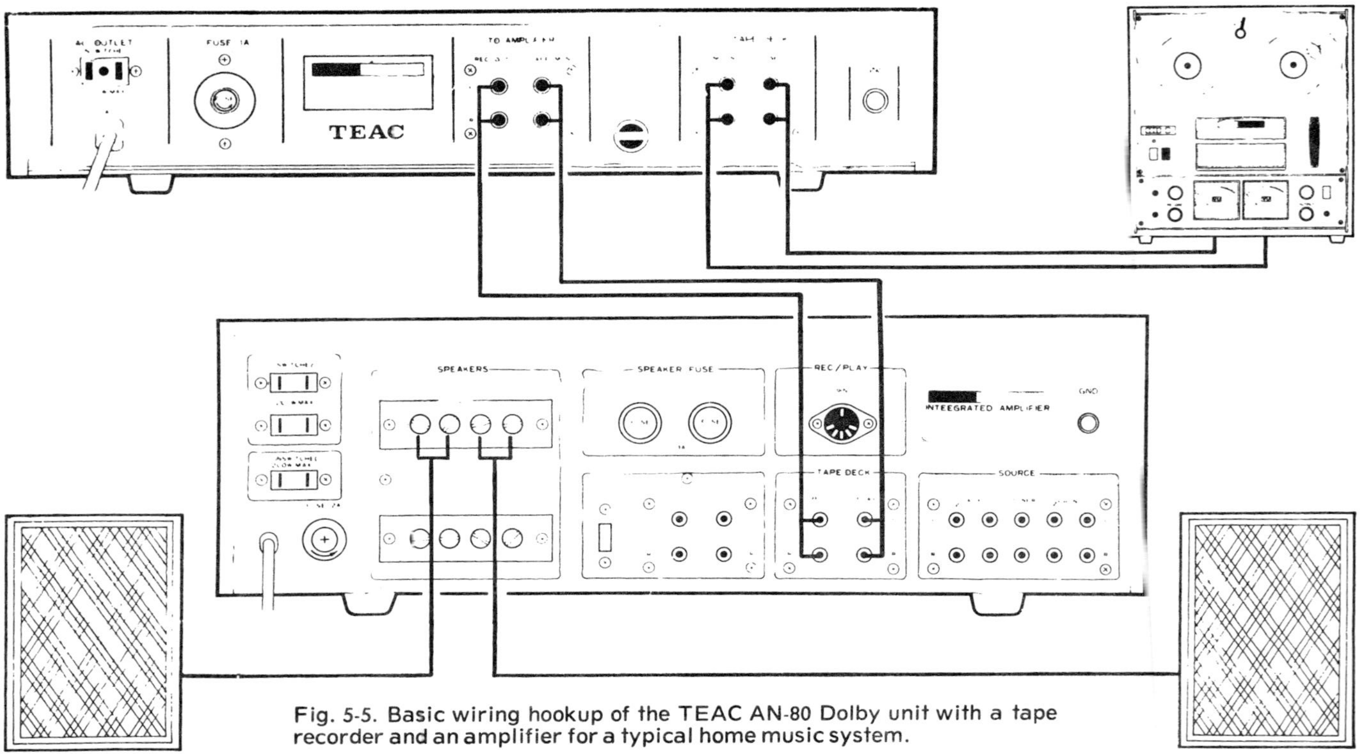

Fig. 5-5. Basic wiring hookup of the TEAC AN-80 Dolby unit with a tape recorder and an amplifier for a typical home music system.

Now that the Dolby B system has been in use for several years, some mass tape duplicators such as Ampex have begun producing cassettes in the Dolbyized format using the B system. The B system uses the highest of the four Dolby frequency bands—the band where hiss occurs. The result is to eliminate the most annoying part of tape's inherent noise, and the reproduced sound comes barreling through in unbelievable quality—for a cassette.

VERSATILE MACHINES

Home systems and equipment come in a variety of types and packages. There are cassette changers that play or record a stack of cassettes: there are machines that will play both sides of the cassette; and there are just plain old ordinary one-sided manually operated machines. Generally, those with the sophisticated electronics are in the last category, since a machine with switchable Dolby, switchable bias, and automatic cassette changing mechanism—all in one unit—would be prohibitively expensive.

Playback-Only Changer

By way of example, a typical playback-only changer, made by Craig (see Fig. 5-7), retails for $189. This machine will play a stack of six cassettes and will play both sides of

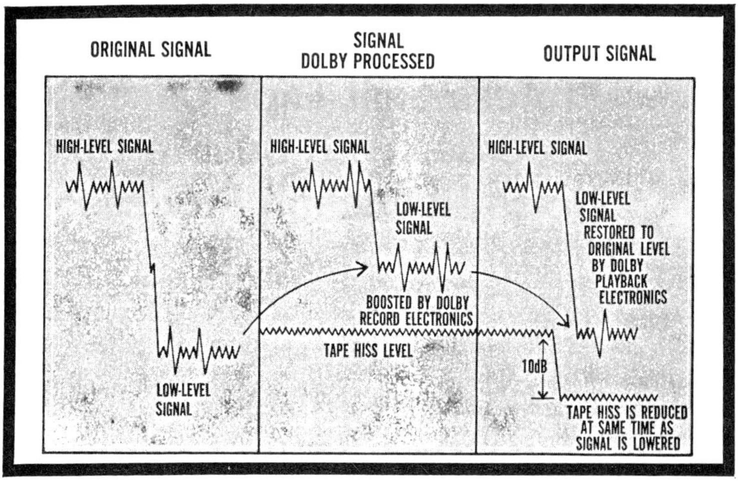

Fig. 5-6. The Dolby boosts low-level signals during recording to bring them above the tape hiss level, then reduces low-level signals in playback. Hiss is reduced by the same amount, pushing unwanted noise down into the mud. (Courtesy Harman-Kardon.)

Fig. 5-7. The first auto-reversing stack-loading cassette deck on the market was the Craig play-only unit. A pushbutton selector lets the user choose between one-side-only or both-sides playback. The unit will not record.

each cassette by reversing automatically at the end of the first side of each. This play-only machine was the first one to hit the market, and it side-stepped the problem of special purpose three-in-one recording heads (which were not available at the time) by not providing the recording function.

Reversing Machines

The problem with reversing machines is that record, playback, and erase functions must all be carried out by one combination head. The reason for this is the configuration of the tape cassette head openings (see Fig. 1-2).

In its original concept, the Philips cassette was designed for manual turnover in a portable machine. The three square holes were designed to admit (1) the capstan pinch roller; (2) the record-playback head; (3) the erase head. Naturally, in an automatic-reversing machine this arrangement is not quite possible, since the number three hole becomes the number one hole during side number two playback and must accommodate an alternate capstan pinch roller. Thus, the middle hole must take all heads, and it took the industry a while to come up with a combination record-playback-erase head that could fit into this slot and still contain all the necessary stereo functions for **both directions** of tape travel. This head is a marvel of modern engineering design and was a long time in coming.

A typical record-playback machine that provides automatic reversing and will operate equally well in both directions is shown in Fig. 5-8. While this particular unit does not offer stack loading and automatic changing, it does have a host of other features including switchable bias. The price tag for all these features is high.

PIANO-KEY CONTROLS

Basic drive mechanisms vary only slightly in home units. The general practice now is to use piano-key type controls and

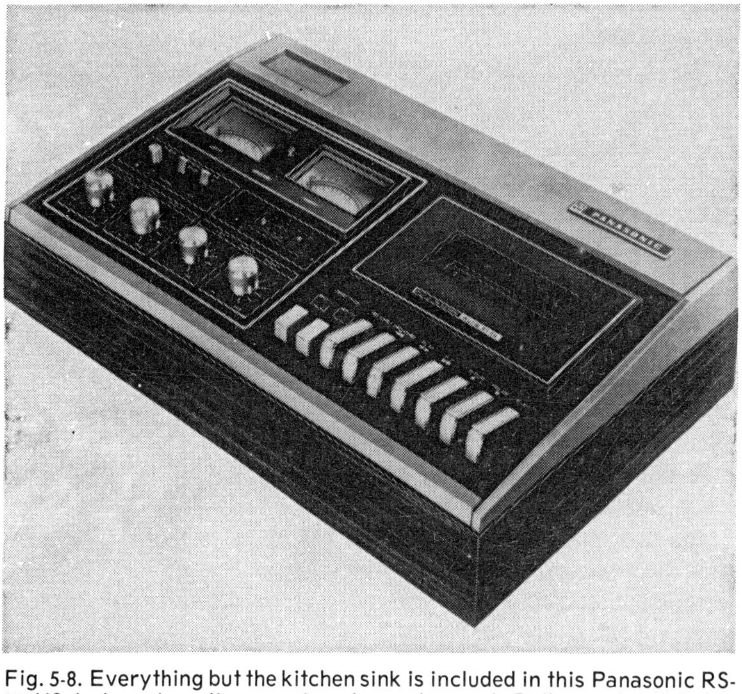

Fig. 5-8. Everything but the kitchen sink is included in this Panasonic RS-272US deck: automatic reversing play and record, Dolby, and switchable bias.

A

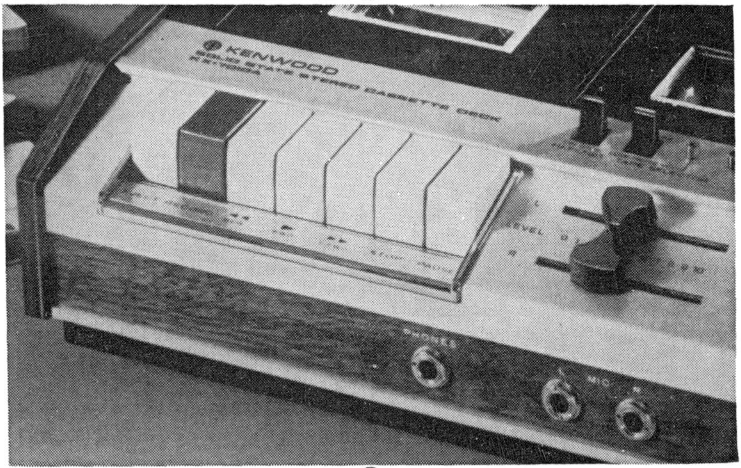

B

Fig. 5-9. Typical of single-lever controls is the play/rewind/fast-forward control on this Panasonic unit (A). More popular and convenient are piano-key controls as shown on this Kenwood KX-7010A (B).

a somewhat complicated lever-action linkage between these controls and the drive system itself. Some home machines still use the classic single-lever control with spring return. While such a control system is certainly much simpler and potentially more trouble-free than the piano-key type, it does little for user convenience. A comparison of what the user is up against is shown in Fig. 5-9.

A crucial feature of any home recording system is an effective pause control. While such controls are usually found only on machines equipped with piano-key pushbottons, even here the types of pause controls can vary. Some companies only offer nonlocking pause controls where the machine will stop running only as long as the key is depressed. To be really useful, a pause control should be capable of being locked in place, holding the tape in the same position with the clutch disengaged for an indefinite period of time.

When the pause control is pressed, the machine's mechanism is not turned off, but simply drawn away from the tape. The tape still remains in intimate contact with the heads, but the capstan and pinch roller are pulled apart from each other so that there is nothing to pull the tape through the mechanism. The pause control also locks the takeup and feed reels so that there is no tension placed on the tape.

The pause control is especially important during almost any type of recording. This is especially true if key elements of the program coming to the machine are to be omitted from the recording. You may be making a tape of a very interesting speaker, but you want to leave out the off-color jokes, or you may want to record just those off-color jokes; you may want to omit periods of laughter, coughing, restlessness, wordlesness, dead-time, and so forth.

Another use for the pause control arises when the cassette is used as a source of background music or sound effects for a special presentation. The cassette can be cued to a specific starting point and held in readiness—running, but with the pause control depressed. At the proper instant the pause control is released and there's the recorded material right on cue!

VU METERS

Virtually all tape recorders today, except a few portables with automatic level controls, use VU meters to show recording level. Old timers in the business will remember the magic-eye (electron-ray) tube that was used as a level indicator on many early tube-type tape recorders. In its heyday the magic eye was probably more accurate and easier to use and certainly cheaper than the VU meter. But the magic-eye level indicator went the way of the vacuum tube. Today's transistor machines use the simple. economical, and accurate VU meter instead. Stereo machines call for two such meters—one each for the left and right channels.

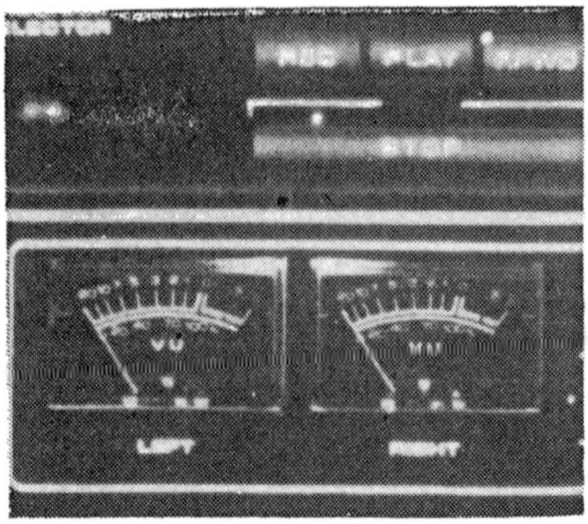

A

B

C

Fig. 5-10. Typical VU meters on stereo recorders (A and B) and on a mono portable (C).

Typically, such a meter has a red area. That's where the meter needle shouldn't go during recording. Some loud transients and peaks might carry the needle into the red area momentarily, but the average highest level should be such that it just about reaches that line. This dividing line is generally indicated as zero VU on the scale. Anything over this will most likely cause distortion in the recording.

At least one tape manufacturer (TDK) disputes this theory of operation. Experts there maintain that their line of premium SD and ED tapes require a higher magnetic saturation level than most other types, and to get optimum dynamic range and signal-to-noise ratio, you should drive the needle into the red on all loud music passages. This naturally depends on the bias-drive setting on your machine. But the proof is in the trying, and—at least for TDK tapes—it works! Before making any priceless recordings this way, experiment a little to find the best VU meter settings for your equipment.

There are all types of VU meters; some of them have no markings at all except for the color bands that we mentioned, while others are elaborately calibrated for a multitude of functions. Some typical VU meters are shown in Fig. 5-10. A few home cassette recorders use only a single VU meter mainly because the manufacturer feels this is more economical. There may or may not be a choice provided for the meter function. If there is no meter function switch, then the meter generally shows the highest level of the two channels being recorded. A selector switch will enable the user to select just the greater of the two-channel function or to show just the left or the right channel. A setup like this is cumbersome, time-consuming, and often causes difficulty in making recordings accurately.

If at all possible, try to purchase a stereo unit with two VU meters for the sake of easy operation. Ideally, these meters should be side-by-side so both of them can be watched at the same time. It seems the height of folly to design a tape recorder with left- and right-channel meters on opposite sides of the equipment. Yes, we have two eyes, but generally they both look in the same direction at the same time.

INPUTS AND OUTPUTS

The typical cassette deck today shows off its sophistication by having far more than the usual complement of input and output jacks. Typically, there may be three pairs of line-input jacks designed for low-level, medium-level and high-level signal inputs. Chances are you'll use the low-level input most of the time since we never seem to be able to get enough signal level out of our available sources. The recorder will also have a pair of jacks called line output to feed a stereo amplifier. There will also be two microphone input jacks—usually located somewhere out front so the microphones can

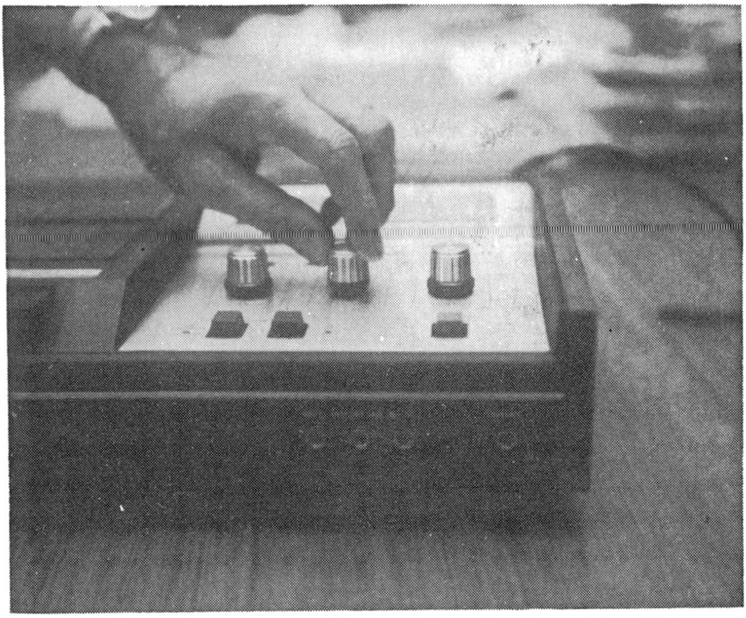

Fig. 5-11. Unusual among cassette recorders, the Sansui SC-700 has an input for a third (center) mic and a separate volume control for it.

be plugged and unplugged easily. At least one deck also accepts a center microphone which can be moved electronically (see Fig. 5-11).

A few units provide a little extra amplification for driving stereo headphones and a stereo headphone jack is located somewhere on the deck, but very often in an inconvenient place.

As of this writing, there are no cassette decks on the market with a built-in Dolby noise-suppression system that offer external inputs and outputs for the Dolby circuitry alone. Such an additional feature would be welcome and certainly a major sales incentive for any equipment providing this function. The object here would be to let the user treat the Dolby circuit as a separate component, plugging it into other tape units or systems that require it. There are none around right now, but possibly there might be by the time you read this.

The deck is the heart of the home cassette system. It can be easily converted into a home music system by the simple addition of an inexpensive stereo amplifier and a pair of bookshelf speakers. Yet, many will find it more convenient to

Fig. 5-12. The Panasonic "Thayer"—cassette stereo recorder, AM/FM stereo receiver and amplifier, and a pair of stereo speakers—is typical of the types of systems that are becoming popular.

purchase a complete system with the cassette unit built into the same cabinet with an amplifier and an FM stereo radio. In such cases, it becomes very difficult to update any portion of the system. Yet, there's a great deal to be said for units such as the one shown in Fig. 5-12. With the rapid rise of the cassette, home systems of this type are becoming increasingly popular. Unfortunately, it also adds to the complexity of keeping this equipment operating properly, and the servicing problems are only made that much more difficult by the equipment's incredible versatility and intricacy.

AUTOMATIC EQUIPMENT CASSETTE

Chapter 6

One of the great convenience features of many open-reel tape recorders on the market today is automatic reversing. This is a system that permits playing of both sides of a tape without manual reel turnover. In the case of extra-long tapes, this feature can provide as much as 12 hours of playing time.

To obtain comparable results with a cassette is difficult at best. The longest cassette on the market today is the C-180, which provides three hours of playing time (90 minutes on each side). Even at that, the C-180 is only available from one manufacturer (TDK), while most other companies provide the C-120 as their longest cassette. This means that if a machine is designed for automatic reversing, the most it can possibly provide is three hours of unattended music playback. A very few machines, such as those shown in Fig. 6-1, have this feature today, and it certainly is a valuable aid and convenience.

STACK LOADERS

One of the earliest attempts to provide long periods of unattended cassette playback was the stack-loading cassette changer. The first one in the series was introduced by Philips several years ago. This type of unit (see Fig. 6-2) held the spotlight as the industry's only changer for quite some time. This unit, and many of those that followed, plays one side only of each cassette in a stack of six—very much like a record changer playing one side only of a stack of discs.

Ideally, a cassette changer should accept not only a stack of six or more cassettes, but should be able to play both sides of each cassette in sequence. This means that a complete album, symphony, Broadway show, or whatever else you're listening to could be heard all the way through instead of half

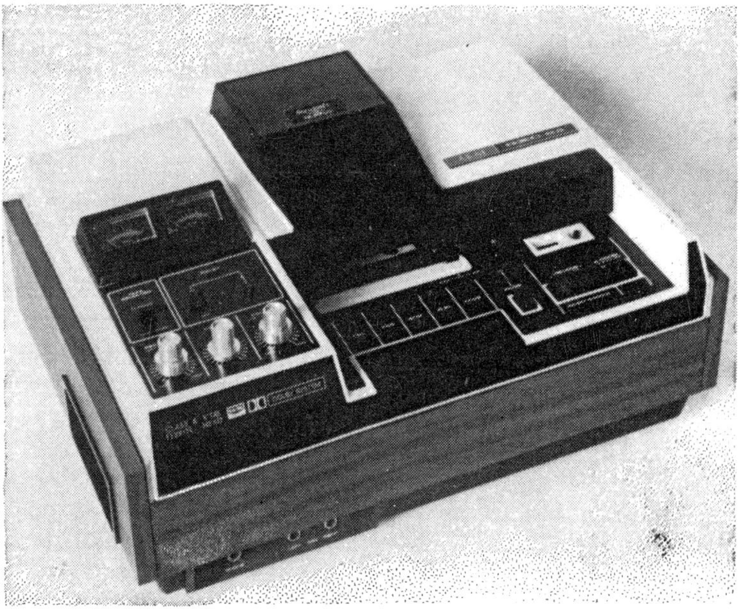

Fig. 6-8. Cassettes ride a carriage atop this 8-cassette reversing changer, the Benjamin RAC-10, which can record or playback continuously.

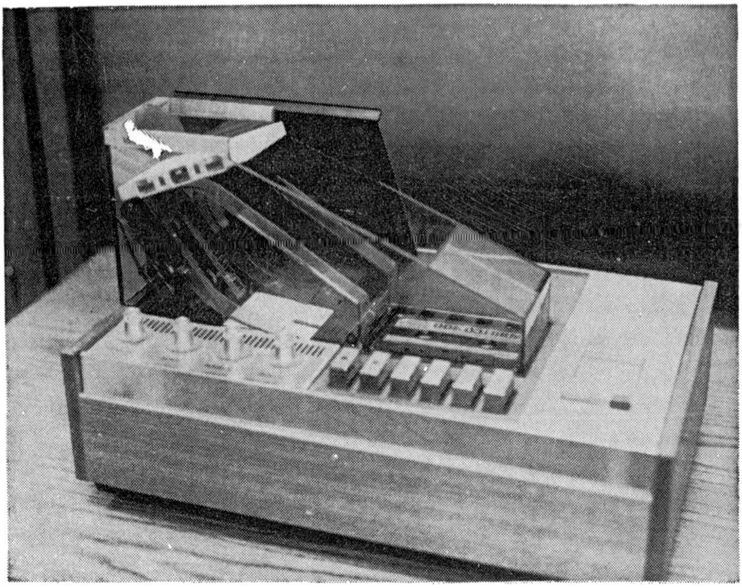

Fig. 6-2. The original stack loader, the Norelco 2401 deck, here fitted with $15 plastic circulator which flips cassettes over for reverse-side and continuous play and play and play.

of it being intermixed with half of each of many other programs. Even in the case of multiple cassettes (operas and other long works now appearing in 2- and 3-cassette groupings) a reversing machine is called for, since these cassettes do not have the kind of record-changer stacking sequence followed by phonograph discs. Sides one and two are on the same cassette, while three and four are on another cassette in the series.

But the mechanical difficulties of designing a changer with automatic reversing—or for that matter designing any kind of automatic reversing machine—seemed to be insurmountable for several years. As shown in Fig. 6-3, the square cut-out holes in the cassette would ordinarily not permit automatic reversing.

Figure 6-3 shows how all three square holes in the cassette are filled by a single-direction recorder: by the rubber pinch roller in the first hole; by the record-playback head in the middle hole (the hole with the pressure pad); and by the erase head in the third hole. Any reversing machine **must** have the use of that third hole for the rubber pinch roller for side B operation.

To make this possible, an entirely new kind of head became necessary a head that would combine the record-playback and erase functions for **four** tracks. It was possible to design such a head, but it took time, and it wasn't until late 1971 that the first reversing record-playback machines made their appearance.

The ingenious design of the pinch roller and head assembly shown in Fig. 6-4A is used in the very first cassette reversing and changing machines—a playback-only unit made by Craig. The reason that the Craig machine beat everyone else's to the market was that it was strictly a playback unit and, therefore, did not require an erase head in that third square hole cutout. Even at that, the single playback head had to be designed for 4-track playback (Fig. 6-4B) with the switching from side A to side B done electronically.

The Craig unit, which was virtually a prototype for reversing changers to follow, is a big, beautiful piece of

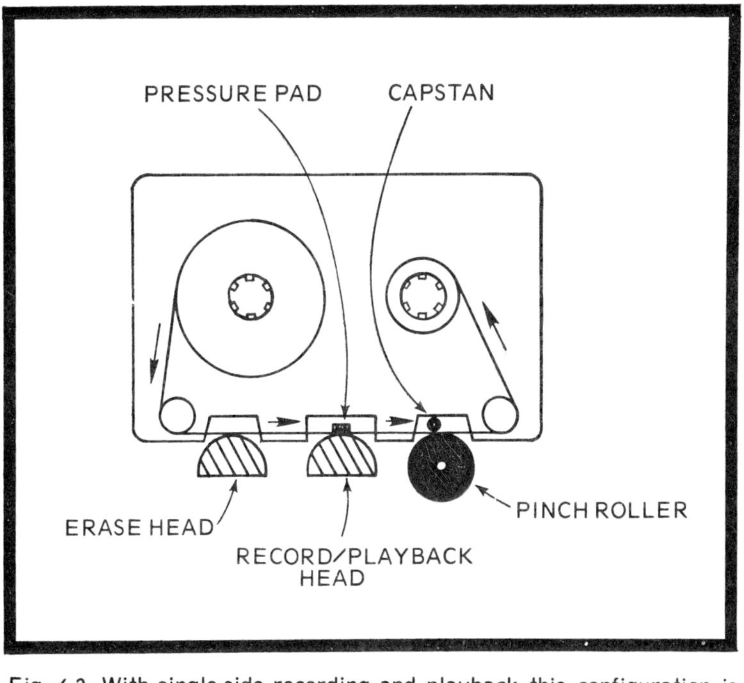

Fig. 6-3. With single-side recording and playback this configuration is used. As you can see, the erase head fills a hole that must accommodate a second pinch roller in reversing machines.

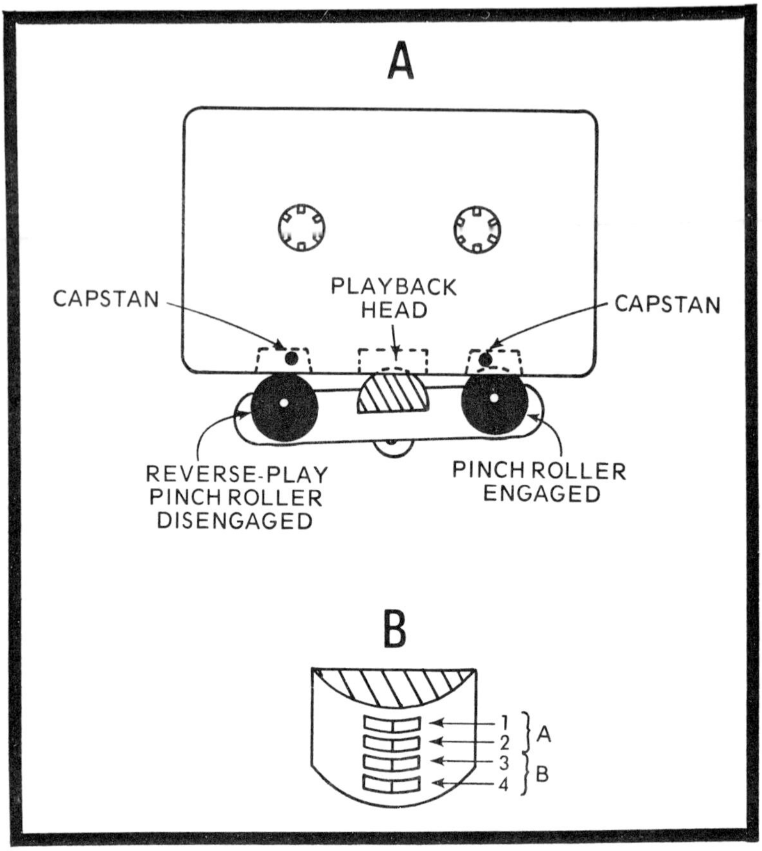

Fig. 6-4. Sketch A shows how the Craig playback machine engages and disengages pinch rollers for reverse playback. In the head used for reverse playback, tracks 1 and 2 are switched in for side A; tracks 3 and 4 for side B.

machinery with very complex insides. Like all cassette changers, it was fairly expensive for a playback-only unit ($189) and has since been replaced by an up dated model.

The need for both-side playback was evident to most manufacturers—even to the extent that Norelco introduced a plastic cassette flipper called the "Circulator" as an add-on accessory for their changers. This $15 item, when added to the top of the changer, takes cassettes as they are rejected and shoves them through a curved trough, flipping them over onto the top of the stack for second-side playback.

The problem with this type of approach is that the second side of each cassette is not played in sequence; instead, the stack is "turned over" automatically. Also, once the second stack is played, the first stack starts over again (since this is a continuous playback operation) until someone turns the machine off.

In the intervening time, other manufacturers—notably Panasonic—have tried their hand at marketing a stack-loading cassette changer for one-side-only playback. The model RS-295US ("Manchester") was Panasonic's entry to this field and a moderate number or units have been sold. Other companies making changers of various kinds include JVC, Qatron, Benjamin, and Concord.

CAROUSEL PLAYER

An unusual concept in automatic playback is a Panasonic 20-cassette carousel (see Fig. 6-5) which permits total automatic playback or cassette selection by pushbutton—something like a home-model jukebox.

Fig. 6-5. Carousel-type cassette changer, the Panasonic RS-296US holds 20 cassettes which can be preselected on the control panel, jukebox fashion.

Fig. 6-6. Closeup of metal-foil sensing tape on the ends of a Sony cassette. Some machines rely on this foil to actuate the reversing mechanism.

As with any automatically triggered tape system, some method must be used to sense the end of the tape to trigger either the automatic reversing or the changing mechanism. The obvious and most commonly used method is a torque-sensing device which is activated by the increased drag on the tape at the end of the playback period. This is a built-in feature of any cassette, since both ends of the tape are locked to their hubs. A less common sensing arrangement uses metal foil pasted onto the oxide side of the tape to trigger an electrical sensing device. Some manufacturers—particularly SONY—rely on this type of sensing system in their automatic cassette machines. Some cassette manufacturers provide tape with the metal foil stuck into place at each end (see Fig. 6-6).

Since only a few tape manufacturers supply cassettes with the metal foil, and since only one or two equipment manufacturers rely on this technique, it may never become standard in the trade the way the torque-sensing mechanism has become. All cassettes will trigger a torque-sensing machine; not so with metal-foil machines.

REVERSING MACHINES

A basic automatic reversing recorder such as the Toyo (see Fig. 6-7) provides a host of special features.

This machine will both record and playback, and it also has a repeat function that, when activated, will rewind the tape after playing the first side and play it over again. Unlike most decks, this one comes equipped with a pair of microphones.

A relatively unusual design in cassette changers is the sliding magazine type of unit made by JVC, Concord, and Benjamin. This type of machine holds the cassettes in a tray which slides horizontally, dropping the cassette into the play or record position and then picking up the finished cassette and moving to the next position. Machines of this type can easily be designed to hold many more than the six cassettes that are a practical limit for stack loaders. The unit shown in Fig. 6-8 will hold up to 12 cassettes, which conceivably could provide as much as 36 hours of playing time when used with a C-180 size, or certainly up to 24 hours with the C-120 cassettes. A machine like this can be turned on and left unattended for a full day or more, and it will provide music all of that time.

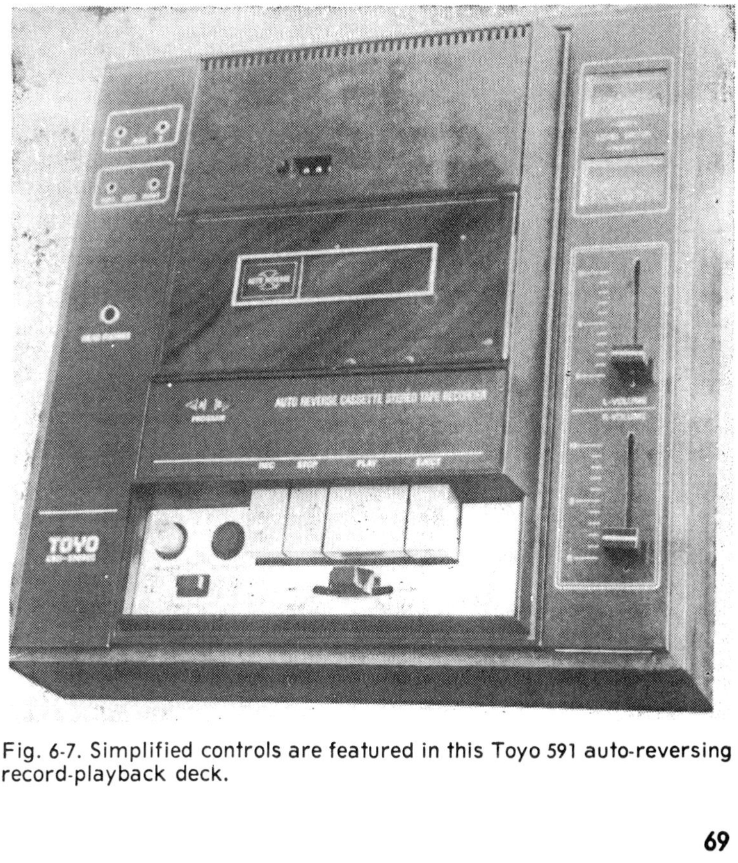

Fig. 6-7. Simplified controls are featured in this Toyo 591 auto-reversing record-playback deck.

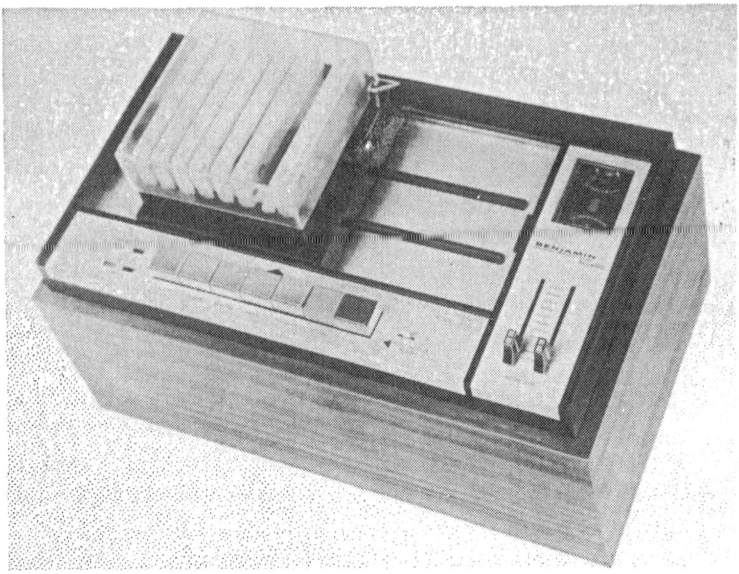

Fig. 6-8. Cassettes ride a carriage atop this 8-cassette reversing changer, the Benjamin RAC-10, which can record or playback continuously.

Generally, a changer requires nothing much more in the way of care than any other fine music instrument. It should be kept covered when not in use to keep its interior dust-free, since dirt is the major enemy of all cassettes. Periodically, the cover of the machine should be removed and all accessible levers and other movable parts and actions should be cleaned and lubricated appropriately. Remember that some types of surfaces require a high-viscosity lubricant, such as Lubriplate, in order to work properly over an extended period of time without binding. Generally, this type of lubricant is good for about 90 percent of working mechanical surfaces in a cassette changer.

When performing such servicing, treat the mechanism like you would that of a record changer—looking for obvious problems, letting it cycle while you watch, and cleaning and oiling wherever you think it necessary. Remember that no lubricant should get on any of the rubber parts of pulley surfaces, since this could cause serious wow problems.

AUTOMOTIVE EQUIPMENT

Chapter 7

Highway listening became a favorite pastime with the introduction of the 8-track cartridge. Now, the cassette is usurping part of the cartridge's natural market, and more than a few cassettes can be found criss-crossing the country on our highways.

There are four basic types of automotive cassette systems. The first is simply a factory-installed cassette-AM-FM radio combination that comes with your new car. Purchased this way, a cassette player costs around $300—quite a chunk of money, but it really pays for itself in convenience and theft protection.

The second type of unit is an add-on cassette player which can be hung on any convenient point under the dashboard or mounted on the floor. There are many types of play-only units. A typical one is shown in Fig. 7-1.

The third type of unit provides not only playback but monophonic recording as well. This is especially handy for an executive on the go who wants to dictate important material while tied up in a traffic jam, or otherwise occupied in commuting or on a business trip. Generally, the microphone clips onto the side of such a unit (Fig. 7-2) and is the first item that is lost or broken.

The fourth way of using cassettes in the car is with an ordinary portable unit, usually monophonic. It is operated either by plugging it into the cigarette lighter socket or by placing it in a special underdash or console mount designed for quick-and-easy removal (see Fig. 7-3). While this method isn't nearly as satisfactory or as satisfying as the built-in or add-on stereo deck, it still can serve its purpose well.

STAAR LOADERS

Many automotive recorders and players use the Staar loading system. Unlike most cassette systems, in which the cassette is completely enclosed in the recorder, the Staar system leaves the tabbed rear of the cassette exposed. The playback mechanism moves up to meet the cassette. How this is done is shown in Fig. 7-4. The Staar system player is easier to load on the road, and it automatically ejects the cassette after it has finished playing. Both of these features are very important for fumble-fingered drivers who obviously can't take both hands off the wheel to operate the tape equipment.

POWER SUPPLY WIRING

The wiring system for cassette recorders is basically simple and straightforward. A ground connection should be made to the automobile chassis from the point indicated on the cassette unit. The power lead should be taken off a line that turns on and off with the ignition. This is an important safeguard; otherwise, the cassette player could be left running accidentally when you're out of the car, and could run down the battery.

A quick and dirty way of connecting a wire for the hot lead is to slip a bare wire end under one of the cartridge fuses in the fuse block as shown in Fig. 7-5. A more permanent connection can be made with clip-on connectors such as those used by most trailer rental agencies.

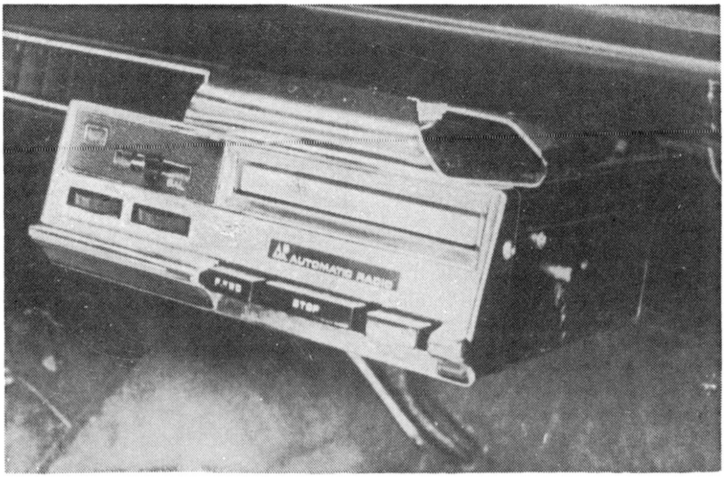

Fig. 7-1. Under-dash mounting of Automatic Radio cassette play-only unit, using a quick-release bracket for easy unit removal.

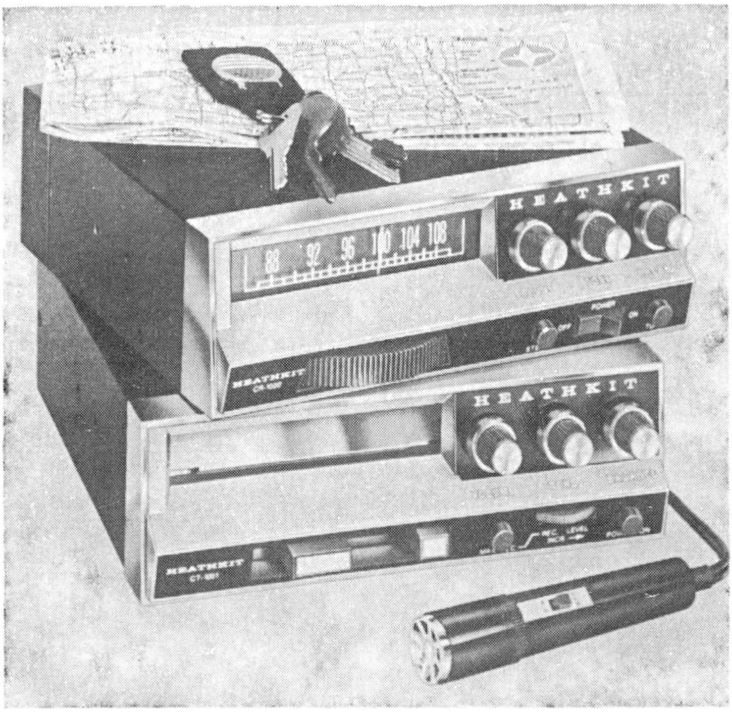

Fig. 7-2. A Heathkit pair, a stereo FM radio and a stereo cassette recorder designed for under-dash mounting.

SPEAKER WIRING

The speaker wiring is relatively simple. The cassette player will have two wires for the two stereo speakers. The common connection for each speaker is the chassis of the car. These wires are usually color-coded and the installation instructions will indicate which color is for which speaker.

When running wires to the speakers mounted flush in the doors, be sure to allow enough slack for opening and closing the door. Allow some extra slack to hang down in a loop inside the car door so that any excess wear can be accommodated. If the wire at the door hinge becomes intermittent or broken from repeated flexing, you can simply pull some of the wire slack out and repair the break with some of this left-over wire.

When running the speaker wires to the rear deck, a little ingenuity and caution can provide a very neat and well-hidden job. Speaker wires will fit nicely under the middle foot

Fig. 7-3. Hump-mount console for Panasonic RS-248 cassette player.

molding on the inside of the door on the passenger side. Generally, the passenger side of the car is preferable for this type of installation, since that's probably where the stereo unit will be located. Besides, there are so many wires on the driver's side that adding more to that tangle will only result in a mess, making it difficult to service properly. To see how wires are run under this molding, see Fig. 7-6.

To get to the trunk, you'll usually have to remove the rear seat, stringing wires into the trunk through any convenient openings, or drilling a small hole if needed. Getting the rear seat back into place and well latched is probably the most difficult part of the installation. This isn't anything that requires a knowledge of electronics; just finding the right slot for the seat latch!

If the rear speakers to be used are the flush-mounting type, be sure that the car already has cutouts in its metal

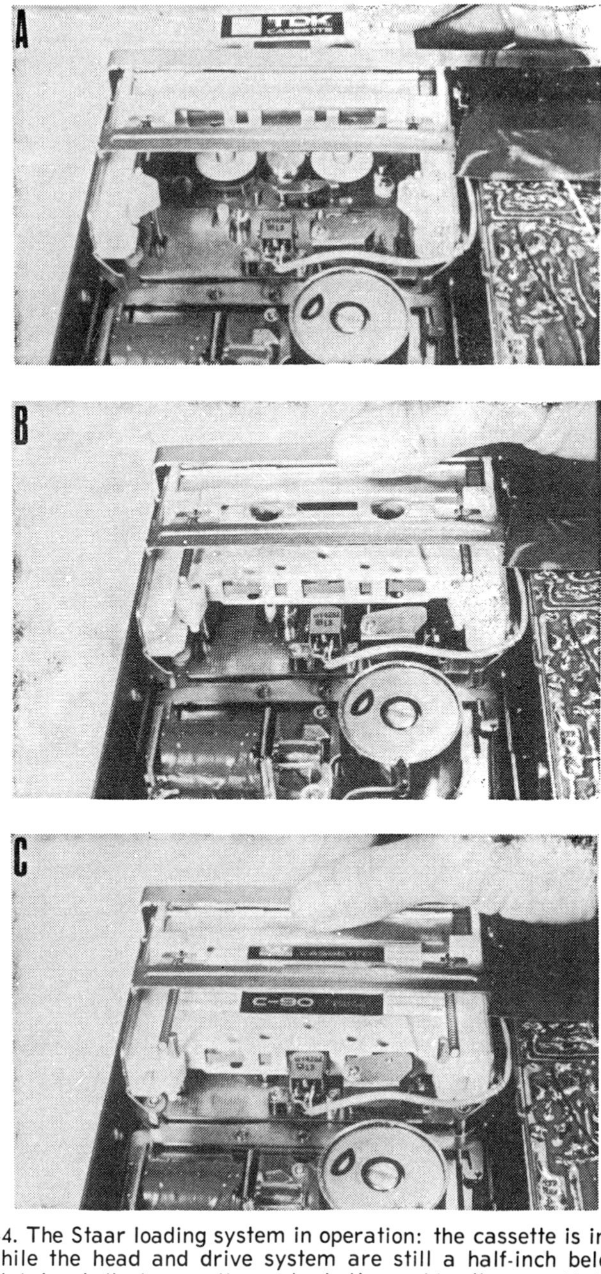

Fig. 7-4. The Staar loading system in operation: the cassette is inserted (A) while the head and drive system are still a half-inch below the cassette's level; the lever action, actuated by pushing the cassette into the player (B), lifts the playback platform to the same level as the head; then the cassette is locked into playback position (C).

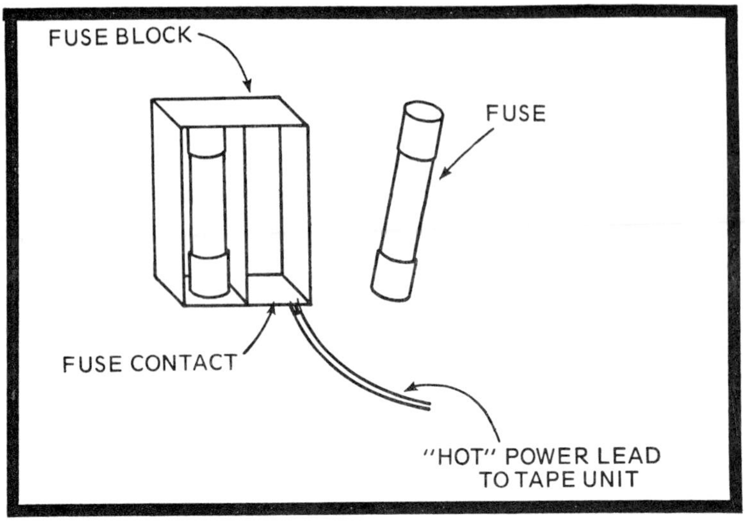

Fig. 7-5. Quick and dirty way to connect the hot lead to your cassette recorder: slip the stripped wire end under the fuse cap. Pressure holds the wire in place.

framework for such speakers. Most new cars have two cutouts for 6 x 9 oval speakers in the rear deck, and the only cutting needed is the opening through the fiber cover on the inside of the rear deck itself. A Stanley knife will do a neat job of cutting out these holes. Be sure to mark off the holes carefully with the speaker mounting grilles used as templates before you go ahead and actually cut. These grilles should overlap the holes, of course.

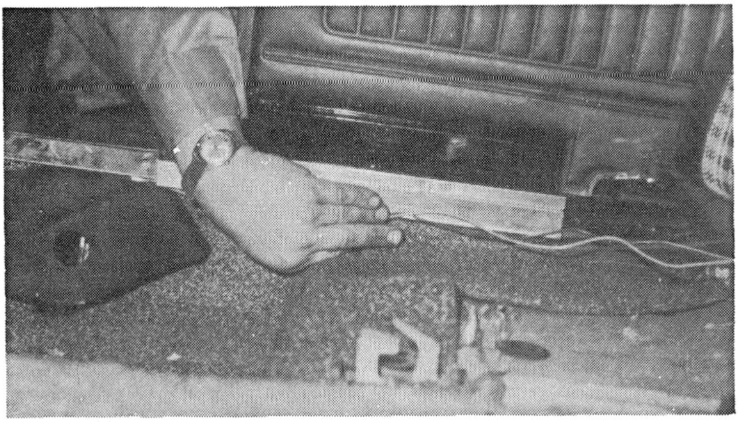

Fig. 7-6. Speaker wires can be run under the door molding trim to the rear seat speakers.

One note of caution: When choosing grilles for the rear speakers, be sure to use **black metal** ones. Plastic grilles have a bad habit of warping in the sunlight and the high temperatures that develop in the car when it sits and bakes all day. Also, a chrome metal grille will reflect in the rear window glass at night, making it difficult to see who's behind you when you're driving. Therefore, use black metal.

When installing those rear speakers, be careful; it's very easy to puncture, tear, or otherwise damage those fragile paper cones. It's a lot more aggravating to replace a speaker because of something that you did wrong than because of a factory defect. After all, you can complain about a factory defect. What can you grouse about if the culprit is you?

The rear speaker connections are fairly straightforward. If the terminals are marked with colored dots or with plus and minus, the plus or the red-dot terminal should be to the "hot" speaker line coming from the front of the car. The negative or black-coded terminal goes to the chassis ground (see Fig. 7-7).

Once this system is installed, there is no reason why it shouldn't work properly under all conditions. If you're doing this for yourself, you'll derive a great deal of pleasure and enjoyment from the cassette stereo in your car. If you're doing

Fig. 7-7. Speaker mounted in the rear deck. If an open-hole grille cover is used, cement a piece of optically opaque speaker grille cloth under the metal grille to prevent sunlight from ruining the speaker.

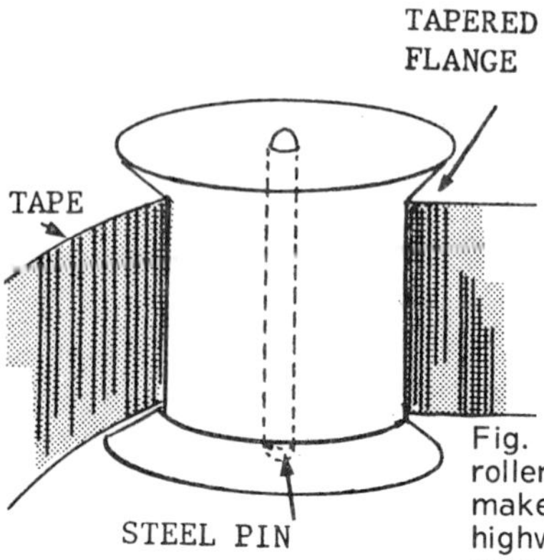

Fig. 7-8. Tapered and flanged rollers used in TDK cassettes make them less susceptible to highway vibration and resulting poor playback.

it for a customer, bear in mind that he may be back with minor complaints or questions about problems with the system.

One basic problem that he may complain about is that every time he goes over a bump the music gets strange or develops a wavering or warbling effect. Explain to your customer that this is not the fault of the system but of the cassette itself and that very few cassettes are capable of sustaining the sharp jolts of the highway without causing them to waver a little bit. While this is generally not a problem with 8-track cartridges, it happens with the majority of cassettes and is possibly one reason why they have not been gaining favor in automobiles as fast as cartridges have. But with this one fault aside, cassettes have so many advantages over cartridges that it's easy to overlook this minor problem.

Certain types of cassettes have better internal construction than others. If the wavering problem persists, advice your customer to try certain premium-grade types that have internal flanged rollers instead of fixed posts. This helps prevent the tape from sliding back and forth when the cassette vibrates. One such tape is made by TDK; the internal construction is shown in Fig. 7-8. A customer still may not be impressed, but at least you've given him an alternative solution to his problem. After that, it's up to him.

AUTOMOTIVE ACCESSORIES AND INSTALLATION

Chapter 8

The American automobile seems to be a natural environment for cassette recorders and players. The compactness and versatility of this equipment and tape make it ideal for highway recording and listening. Certainly it's a simple matter to plop a portable cassette recorder onto the front seat and use it for dictation, note taking, and various and sundry other recording features. It's equally easy to use the same machine for playing back notes, for reminders of things to do, travel directions, or simply to listen to entertainment material.

But if the cassette is going to be used mainly for entertainment in the car, it's more prudent and feasible to have a permanent installation with good-quality door- and deck-mounted loudspeakers for better sound reproduction, as well as a permanent connection to the car's battery. One of the earliest attempts at putting the cassette in the car was a special floor mounting kit for monophonic recorders (see Fig. 8-1).

Soon after the advent of stereo cassette format, playback-only and stereo-playback-with-mono-record machines started to appear on the market. It was tough going for these machines at first, especially since there was very little prerecorded musical material on stereo cassettes at the time. The cartridge was still the preferred medium for highway listening, and it was several years before Detroit started offering cassette equipment as a factory option in new cars. Today, it's possible to buy a cassette player built into almost any new car made in the United States.

As with any type of installation, certain key accessories can make automotive cassette equipment more versatile and more enjoyable. One of the most interesting of these is a cassette-size tuner which slips into the machine and converts it into a radio receiver (see Fig. 8-2). Usually, this type of

Fig. 8-1. Locking floor mount by Panasonic holds a mono or stereo unit, but it is most suitable for compact battery portables.

tuner can be used in portable cassette machines, and represents a minor marvel in miniaturization and electronic design.

SERVICING ACCESSORIES

Servicing accessories are almost exactly the same as those used for nonautomotive cassette equipment. For head cleaning, the usual cotton swabs dipped in head-cleaning solution or a head-cleaning tape are suitable. But when it comes to demagnetizing the heads, we're faced with a different kind of problem.

Home-type head demagnetizers rely on ac house current to provide the field needed to do the job. Just plugging an electromagnetic coil into a car cigarette lighter won't work, since the car's battery provides dc. A specially designed automotive head demagnetizer will work from the car's 12V cigarette lighter socket, but none are yet available for cassettes.

The unit contains a small multivibrator circuit which converts the dc into an ac, which then drives the degausser electromagnet and pole piece. These devices are long and

Fig. 8-2. Portable cassette recorder with an AM radio cassette that slips into the recorder, thus turning the machine into a radio. (Courtesy Bell & Howell.)

slender to make it easier for them to reach into the guts of cartridge players. Unfortunately, most of them are not skinny enough to fit into a cassette recorder slot, since they're designed mainly for cartridge units.

One possible solution is to use a combination head-cleaning and -demagnetizing tape of the kind produced by Ampex (see Fig. 8-3). The tape applies dc demagnetization, which is not as effective as the alternating-current type, but it's better than nothing. A permanently magnetized metal foil tape is drawn across the heads, and it removes much of the head's residual magnetism.

INSTALLING AUTOMOTIVE UNITS

Installing cassette equipment in cars can create several problems. Typically, the manufacturer provides a minimal mounting bracket which allows the unit to be hung under the dashboard or mounted on the floor. The trouble with all such mounts (as any car owner will tell you) is that the tape recorder can be stolen almost effortlessly. There is nothing quite so inviting to a thief as a piece of valuable electronic equipment mounted loosely under the dashboard.

There are several ways of getting around this, and all of them rely on certain types of antitheft devices. Probably the simplest and most universal today is to use a lock mount of the type shown in Fig. 8-4. While this mount contains a key-operated lock, the lock is virtually useless, since it will spring open for anyone who happens to have a hairpin or a paperclip handy. The main value of this mount is the fact that the recorder can be removed easily and put in the trunk of the car for safekeeping. While this is a minor inconvenience, it's a lot cheaper than replacing a stolen recorder every other week.

If the brute force route is more appealing, it's possible to mount the machine more or less permanently in the car with locking bolts and bolt covers of the type shown in Fig. 8-5. However, thieves are very ingenious these days, and even these locking bolts will give way to a determined effort. The chief benefit of security devices of this kind is to slow the thief down and increase the possibility of his being discovered at work.

An alternate kind of lock mount is shown in Fig. 8-6. The bracket is bolted to the dashboard underside. The recorder slides into the bracket and is key-locked into place. The body

Fig. 8-3. Combination head-cleaning and degaussing cassette by Ampex applies dc demagnetization.

of the machine covers the bolts so they can't be removed. The lock is a 5- or 6-pin tumbler, which will resist the picking efforts of most casual parking-lot thieves. But as with any kind of security device, a really determined onslaught on a deserted street might still result in success for the recorder snatcher.

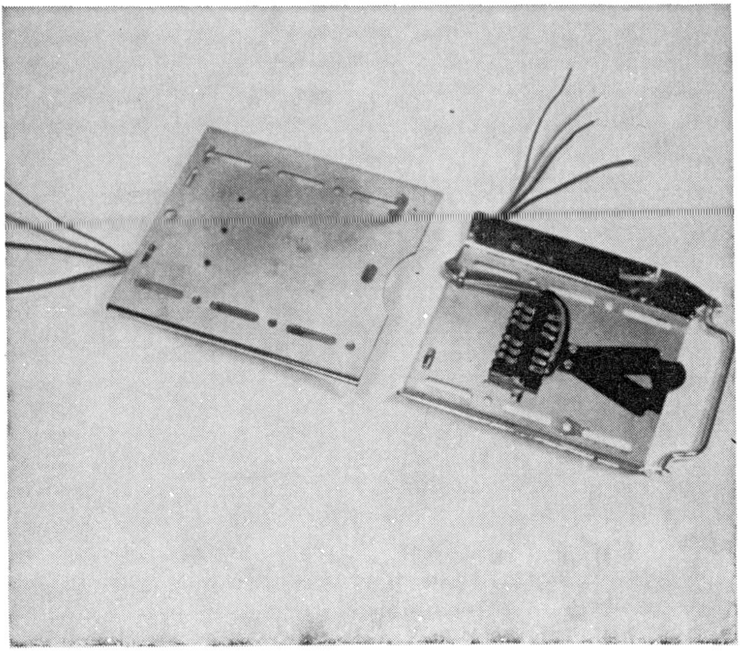

Fig. 8-4. With a quick-release under-dash mount, you can put the cassette recorder in the trunk when it is not in use to prevent theft. This unit is made by Metra Electronics.

BURGLAR ALARMS

Another method that will help to slow the burglar down is the use of various types of alarm devices. One such device is designed specifically to protect electronic gear in the car. Whenever the equipment is touched or moved, the horn starts blowing intermittently, and presumably this will bring the owner of the car running or at least call attention to the car so that the thief will not continue at his work. Another type of alarm system, shown in Fig. 8-7, rings an enormous bell when anyone tries to enter the car illegally.

SPEAKER INSTALLATION

Installing loudspeakers is relatively easy, but it can be time-consuming. The simplest type of installation is to mount a pair of oval speakers in the rear deck of the car. This is a logical location, since the metal frame of the car—at least on most recent models—has cutouts in the metal-chassis understructure so that it's just a question of cutting the oval holes in the fiberboard material of the rear-window deck.

On most medium- to large-size cars, these holes will take 6 x 9 speakers. It's generally a good idea to use the best speakers available, including coaxial units that provide extended-range high-frequency response.

Many of these newer speakers are multi-impedance units with either dual coils or tapped voice coils. Suddenly, you're faced with four speaker terminals to connect instead of two, and the way you wire these up to each other and to the cassette player determines the speaker impedance. The different arrangements for low-, medium, and high-impedance wiring for a typical speaker are shown in Fig. 8-8.

Running the wires to the rear-deck speakers is not a great problem. They can be well hidden under the door molding of the car and passed behind the rear seat. Typically, the most difficult problem comes in replacing the rear seat and getting it to lock into place.

Surface-mounted speakers are available in a variety of shapes and sizes as shown in Fig. 8-9. For rear-deck mounting,

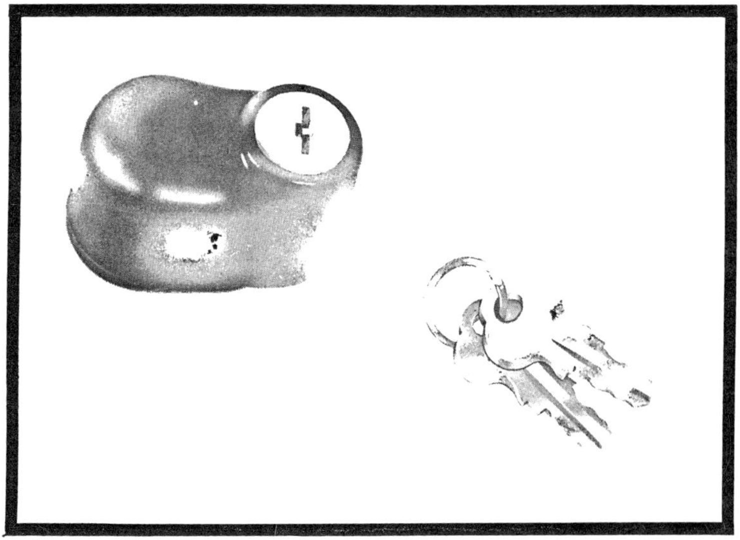

Fig. 8-5. Bolt lock by Audiotex covers mounting bolts and discourages removal.

Fig. 8-6. Under-dash lock-in bracket by Panasonic will also discourage but not totally prevent theft by a determined stereo burglar.

these are certainly a big time saver, since they don't require large oval cutout holes. The problem with them is that they do tend to obstruct the view a little bit. Depending on how large a rear window you have and how large a speaker you decide to use, you will probably opt for a flush-mounting oval speaker rather than the surface-mounting type.

The front doors are the most common location for front-seat speakers, but these can be a little bit difficult to install at times. Speakers of this type are shown in Fig. 8-10. Generally, the door-mount speaker will provide an added dimension to your listening pleasure in the car, but it really should be used only if you're a stereo listener and expect to spend a lot of time playing tapes on the highway.

SWITCHES AND FADERS

An important group of automotive accessories includes speaker switches and faders. Just about any rear-window deck-mounted speakers can also be used as part of your radio playback system with the addition of a simple switch or two. A basic switching control circuit is shown in Fig. 8-11. What this does is to substitute the rear speakers for the single front

Fig. 8-7. The Audiotex burglar alarm makes a lot of noise when the car door is opened by unauthorized persons.

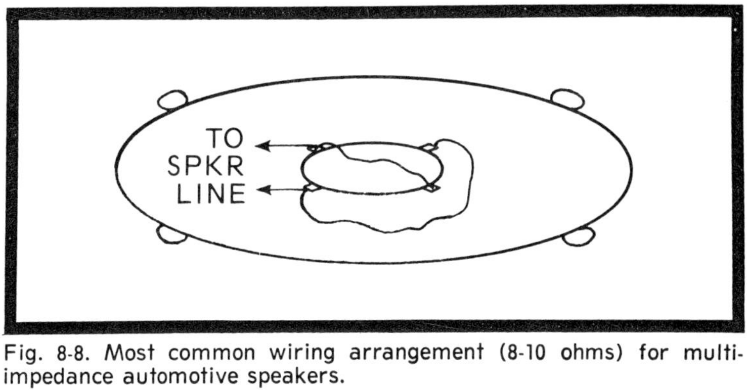

Fig. 8-8. Most common wiring arrangement (8-10 ohms) for multi-impedance automotive speakers.

Fig. 8-9. For fast and easy rear-deck installations, wedge speakers like this one from Metra Electronics fill the bill nicely.

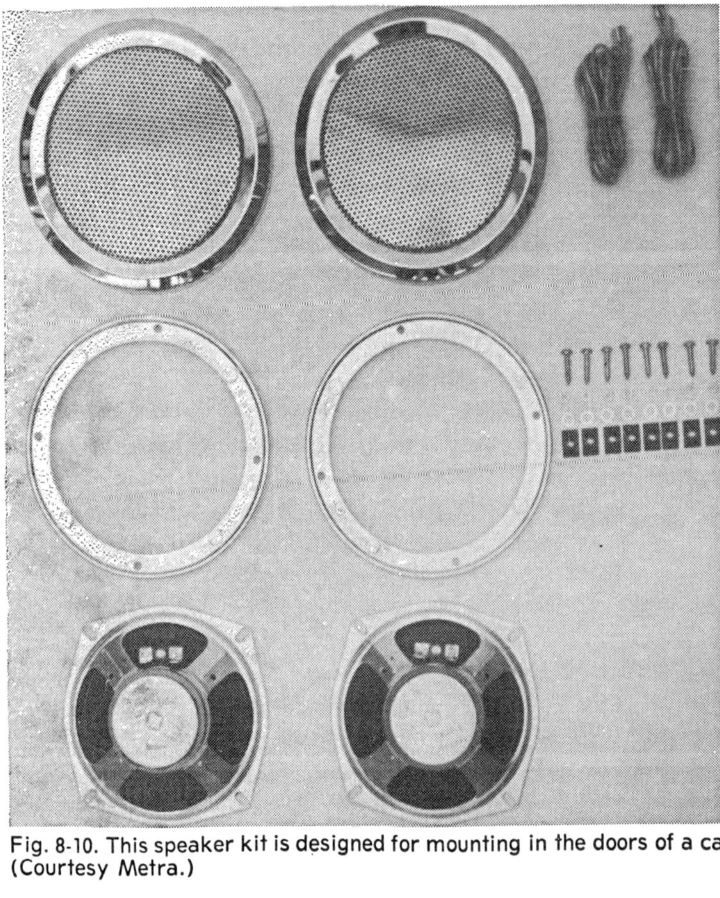

Fig. 8-10. This speaker kit is designed for mounting in the doors of a car. (Courtesy Metra.)

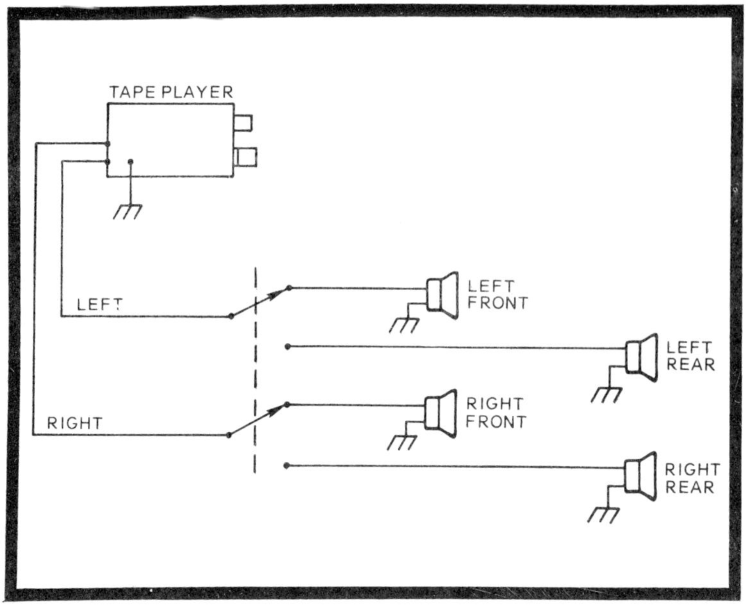

Fig. 8-11. Simple speaker switching system selects either front or rear speakers.

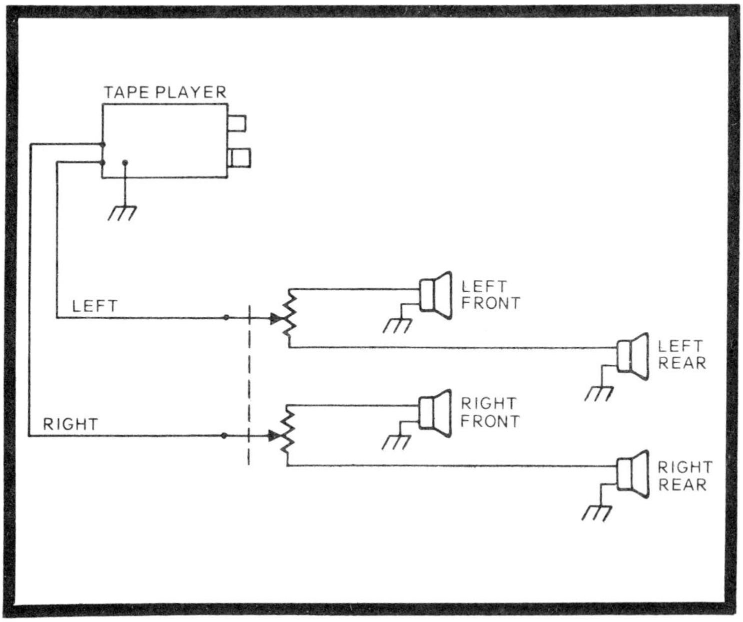

Fig. 8-12. Fader system used to vary the signal applied to front and rear speakers.

Fig. 8-13. "Switch-O-Matic" by Audiotex slips out of sight behind the dash. It plays regular mono radio through the stereo speakers with no special switching needed. Unit supplies its own speaker power.

speaker for conventional monophonic AM and FM radios. A more convenient method is the arrangement shown in Fig. 8-12, which uses just one of the rear speakers mounted on a fader control, which balances between front and rear.

There are numerous commercial fader controls on the market that operate both in mono and stereo. Probably the most ingenious of the available devices for rear-speaker switching is GC's Audiotex "Switch-O-Matic" (see Fig. 8-13). This little solid-state unit is connected between the radio and its speaker and the cassette player and its speakers. What it does is to provide the proper impedance match and power output to drive the stereo speakers from the radio, and it switches over automatically from one to the other. It's an interesting little item housed in a gray Minibox, hidden easily behind the dashboard.

There are many other accessories that cross over between home-portable-automotive equipment. A little experience and ingenuity will certainly show you which ones are most appropriate for use in your car.

HOME EQUIPMENT ACCESSORIES

Chapter 9

Along with the advent of cassette recorders and the resulting burgeoning market, various manufacturers have developed a line of special purpose accessories for using and maintaining home cassette equipment. Some of these items—something as simple as cottom swabs and head-cleaning fluid—are precisely the same as those intended for open-reel equipment. But the heads on a cassette recorder are often difficult to get at, and the problem of cleaning the capstan and rubber pinch roller is rather difficult, too.

The conventional way to clean heads on all tape recorders is with a cotton swab dipped in a strong alcohol solution or a prepared head-cleaning solution. It's possible to clean the heads on many cassette recorders using this method (see Fig. 9-1) by putting the machine into the play mode, which moves the heads out from under the protective cover to make them accessible. The trouble is the heads are still not very easy to get at in many machines—especially the automatic types and changers. Also, if the machine uses a Staar loading system, the heads cannot be reached without at least taking off the machine's cover.

Obviously, the cotton-swab method leaves a great deal to be desired, especially since it does not readily permit access to the capstan and the pinch roller. Another method is to use a head-cleaning tape which can be purchased at just about any dealer.

A distinction should be made between head-cleaning and head-polishing tapes. Very often the latter type carries the head-cleaning tape label, but if you're not aware of the difference it can spell disaster for your tape recorder.

IMPREGNATED TAPES

The most common type of head-cleaning tape uses a cloth that is chemically impregnated (see Fig. 9-2). The chemical

Fig. 9-1. A cotton swab dipped into a head-cleaning solution is the best and easiest way to clean recorder heads, provided they're accessible.

detergent solution in such a tape is good for perhaps six or seven passes through a dirty tape recorder before it becomes exhausted. Some manufacturers include a bottle of cleaning fluid with the tape for renewing the chemical treatment each time the cleaning tape is used. Cassette cleaners of this type will have a much longer useful life span.

A head-polishing tape, on the other hand, should be used very sparingly. This type of tape contains the same abrasive material that is used to polish the heads in the factory before shipping. As with any abrasive, the head-polishing tape grinds a certain amount of metal off the heads and restores to new condition a surface that is excessively dirty, corroded, pitted, and clogged with debris—heads that cannot be cleaned satisfactorily by conventional cleaning methods.

Head-polishing tape coatings generally contain chromium trioxide as the abrasive material, and this in itself poses a certain hazard. When using a chromium-trioxide tape a dust or powder is released in the air. If this dust is inhaled with any amount of regularity, it can cause nasal inflammation and ulcers, leading to a perforation in the nasal cavity. Workers who are constantly exposed to chromium trioxide dust soon develop perforated ulcers in their nasal cavities, causing permanent damage to their breathing capability. Just a word to the wise.

HEAD DEMAGNETIZERS

Another important item in the shopping basket is a head demagnetizer, also known as a degausser. These devices are as absolute must and provide very low-cost insurance against damaging your tapes. A typical degausser is shown in Fig. 9-3.

Fig. 9-2. A chemically impregnated head-cleaning cassette tape is the only way to reach hard-to-get-to heads and guide posts, but such cassettes can only be used a few times before they become saturated with oxide.

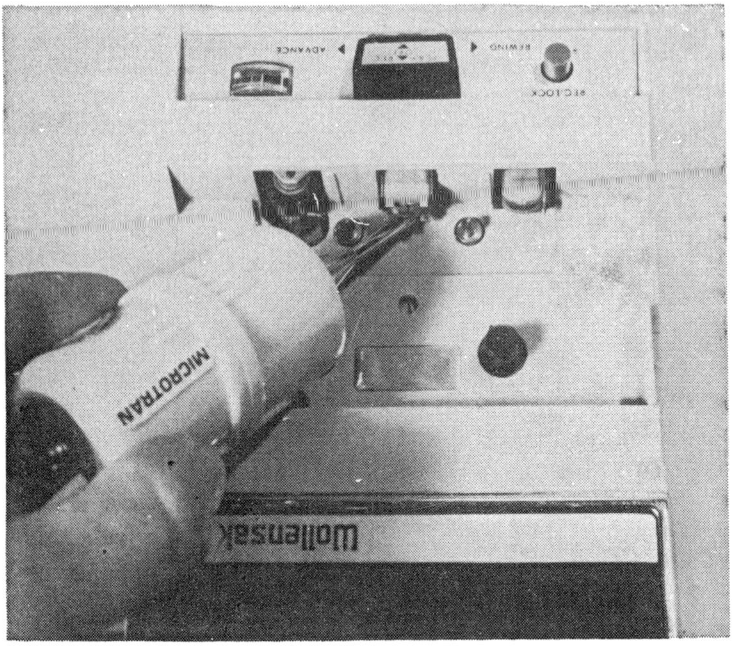

Fig. 9-3. A head demagnetizer is a must. It removes accumulated residual magnetism on the recorder heads. The Microtran unit shown here operates from the ac line.

Most degaussers come with a very nicely polished metal tip, which can damage your tape recorder's heads if it is allowed to touch the head during use. A sure way to avoid this is to coat the tip with a soft plastic or clear nail polish. You can do this by dipping the metal pole piece into one of the many prepared liquid plastics on the market that are available in hobby shops.

The heads in your tape recorder should be demagnetized fairly often, depending on how much you use the equipment. A good rule of thumb to follow is a degaussing for every 12 to 14 hours of playing time. Naturally, the heads should be cleaned at the same time and at least twice as often as you use the degausser.

BULK ERASERS

Another type of demagnetizing unit is the bulk eraser. Generally, these units come in a convenient hand-held model

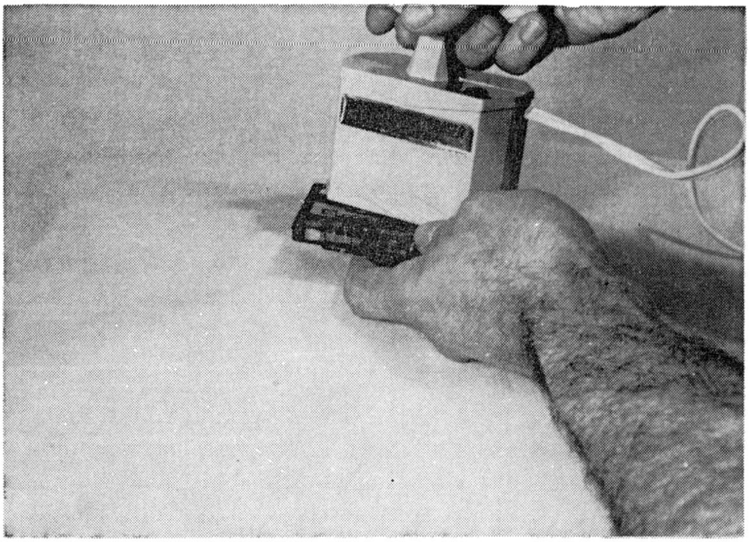

A

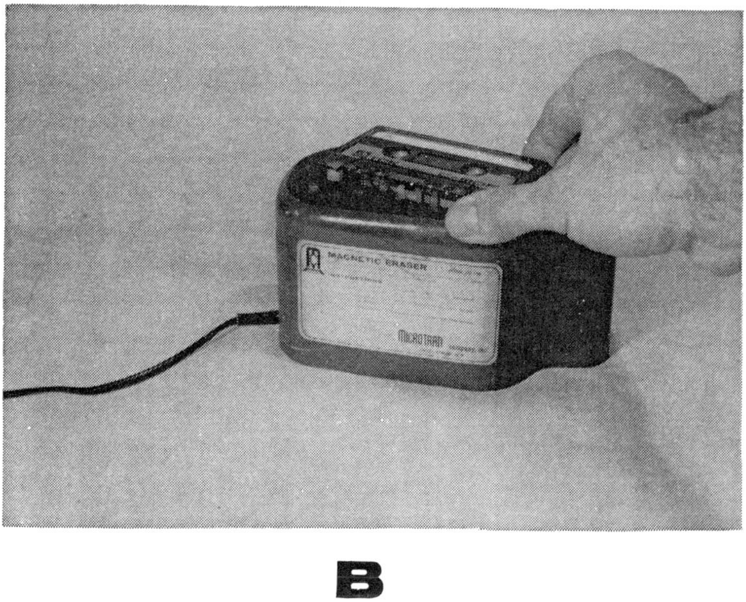

B

Fig. 9-4. The hand-held Microtran bulk eraser (A) is called for when a cassette tape has been reused many times. Bulk erasing restores tape to the virgin magnetic condition. The table model bulk eraser by Microtran (B) is really meant for open-reel tapes. Used with cassettes, it does the job, but is a little bigger than necessary for the job.

(see Fig. 9-4A) or in a larger tabletop size (see Fig. 9-4B). The advantage of the bulk eraser is that it wipes the tape completely clean—removing any residual material or background noise that might be left over from many reuses. It erases the tape uniformly to a factory-fresh condition as far as its noise level is concerned.

One important note of caution: Bulk erasers should never be used near other tapes that you want to preserve; it's also a good idea to take off your wrist watch and park it in a safe place at a safe distance from the eraser before turning it on.

SPLICERS

An aggravating prospect is that of splicing a cassette tape. The tape is so narrow and thin that even getting hold of it to work with it can be a serious problem. First on the list of necessary tape-splicing accessories is a splicing block (see Fig. 9-5). This should be used with a very sharp, single-edge razor blade for cutting. The splice is made with a special gum-backed tape that has expansion properties very close to that of recording tape. In emergency situations with open-reel tape in the past, you may have been able to use conventional cellophane tape for this, but it just won't work with a cassette—even as a temporary emergency measure.

Making the actual splice is extremely difficult because the tape is so hard to handle. It requires a great deal of patience to draw the broken ends together properly on the splicing block and then to make the required cut and splice. It's such a difficult project that for low-cost cassettes, you might be better advised just to junk the cassette and start fresh, unless there is some very important material recorded on the tape you want to preserve.

STORING CASSETTES

Another problem with cassettes is storage. They are small; if you attempt to store them in library fashion, they take up too much space on the book shelf. They tend to collect in and clutter drawers, nooks, and crannies and any other space available. Cassettes can easily become ubiquitous—spreading out all over the house to every accessible location, making it impossible to find the one that you want at a specific time.

There are a number of storage systems available, any one of which may be good for your purposes. A very convenient method is the inexpensive plastic library-type box (see Fig. 9-6A) which stores 24 cassettes in a very compact space. This box folds up nicely and fits on a standard bookshelf, and its handles unfold for carrying.

Another type of case is shown in Fig. 9-6B. While this isn't as compact as the one shown in Fig. 9-6A, it allows you to keep the cassette in its own container, so cassettes may be changed around and swapped easily without having to worry about labeling and empty box storage. This one is especially useful for highway listening, since it keeps the tapes easily at hand in the car.

There are any number of cassette storage systems available for the bookshelf. There's the carousel type shown in Fig. 9-7. There are wooden bookshelf units, cardboard slipcase stackers, many of which are available by mail order or in your

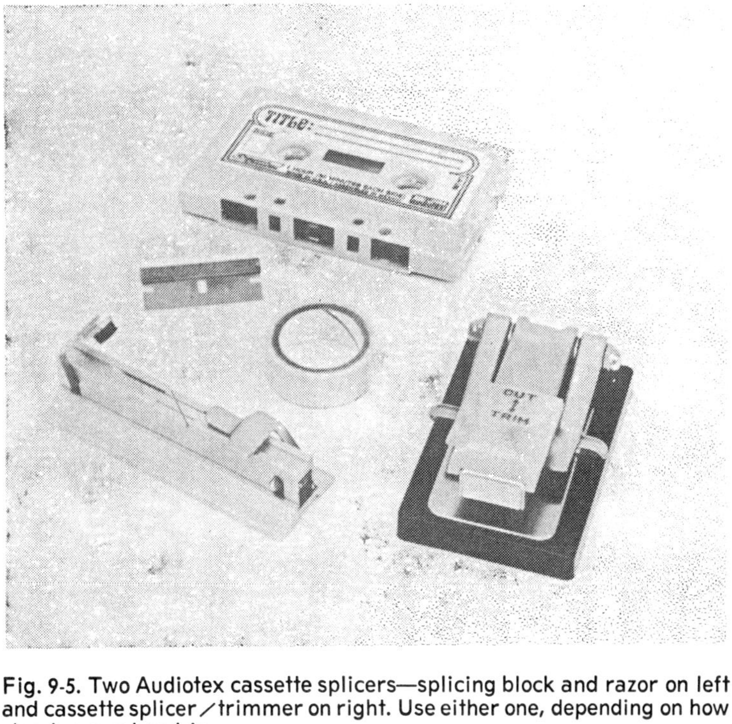

Fig. 9-5. Two Audiotex cassette splicers—splicing block and razor on left and cassette splicer/trimmer on right. Use either one, depending on how steady your hand is.

A

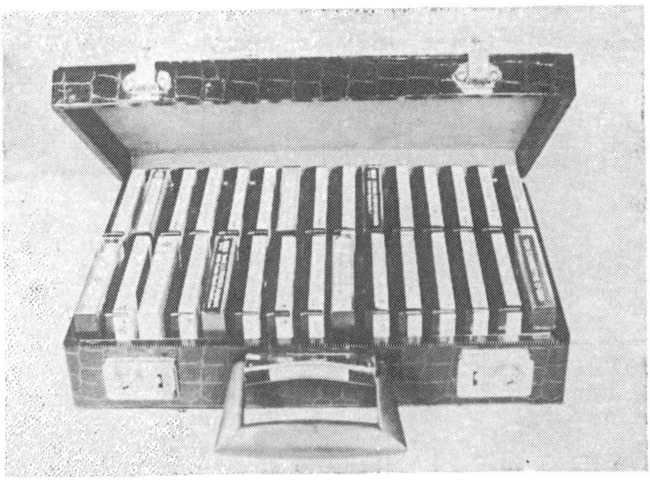

B

Fig. 9-6. Two portable cassette carry-storage cases: The plastic library-type case (A) sells for about $4; it holds 24 indexed cassettes compactly without protective cases: The felt-lined travel case (B) holds 30 cassettes in their protective boxes and costs about $10.

favorite department store's record department. The minute you get involved with cassettes, storage becomes a problem, so try to plan for it in advance.

Marking the cassettes properly is important. There are several methods that can be used, but probably the most convenient is a stick-on label made by Avery, Dennison, and other major stationery manufacturers. These are available from almost any stationery supply store at a very low price; an average supply for a year's use will probably cost a dollar or less.

Markers are important not only on the cassettes themselves, but also on the library boxes and anyplace else where some kind of identification is needed. It's also important to update the label when you reuse cassettes. You can do this simply by peeling off the old label and sticking on a new one. Do not neglect labeling cassettes; an unlabeled cassette can become very elusive when you can't identify it immediately or if you try to rely on memory.

Fig. 9-7. With this carousel-type cassette storage unit, cassettes can be stacked for easy access on a library shelf.

CONNECTING CABLES

Probably one of the most important categories of cassette accessories is the assortment of connecting cables that usually accumulate. These come in a variety of shapes and sizes and the plug types and sizes are covered in the next chapter. One of the things to be aware of is the fact that these connecting cables can become defective very easily and are probably one of the major sources of cassette equipment problems. Wherever possible, use factory-molded cables and connectors rather than home-made types. The home-made type may be called for if you can't get what you need in the store, but factory-molded cables have a longer lifetime and will take a lot more flexing.

Chapter 10

ROUTINE MAINTENANCE

Probably the most common single cause of problems with cassette tape recorders is a lack of proper maintenance at regular intervals. Such maintenance is supposedly performed by the user, but very often is so neglected that a machine ends up in the repair shop for what should have been handled on a weekly basis.

Needing the most attention are the tape heads. As with all tape equipment, these heads eventually become clogged with worn-off particles of oxide and also tend to accumulate residual magnetism. Dirty heads cause a loss of frequency response and can decrease recording and playback signal levels because of the lack of the needed intimate tape-to-head contact. Eventually, a truly bad clogging situation can make the tape recorder almost entirely inoperable.

CLEANING TAPE HEADS

Cleaning the heads can be a very simple matter if they are readily accessible from the outside of the machine. To clean heads, simply dip a cotton swab (see Fig. 10-1) in isopropyl alcohol or in a special tape-head-cleaning solution. The soaked cotton swab rubbed gently over the tape heads will come away dirty with a coating of oxide. Very often it may take two or three such wet swabs to remove all the oxide that has accumulated.

Unfortunately, in many cassette machines, the heads are not readily accessible for cleaning this way, so another method must be used. As shown in Fig. 10-2, a tape-head cleaning cassette will do the job, although sometimes it is not as effective as the old-fashioned alcohol swab. There are several types of such head cleaners; the most common is an impregnated cloth-like material which can be used a few

Fig. 10-1. Heads are sometimes hard to get to with a cleaning swab, as in this Automatic Radio car player. The case was removed here for cleaning and other routine maintenance.

times, and then must be cleaned or discarded. Another type calls for impregnating a cloth tape with a special head cleaning solution each time it is used. One special purpose head maintenance tape made by Ampex also includes a metal magnetic strip which demagnetizes the heads at the same time that it cleans them (see Fig. 8-3).

DEMAGNETIZING THE HEADS

When it comes to demagnetizing the heads, the usual method is to use an electric degausser (demagnetizer). When used at home or in the shop, such a degausser operates from the ac power line. Many such devices turn on the instant they are plugged in and, therefore, should be kept away from any recorded tape material, since the magnetic field surrounding these devices can damage the tapes. Some such degaussers have a built-in pushbutton switch, which is an excellent convenience feature.

To use the device, turn it on by plugging it in or pressing thc button and bring the tip close to—but not touching—the tape head itself (see Fig. 10-3). Preventing the degausser's metal pole piece from touching the tape head is easier said than done, and if it strikes the head in just the right way it can damage it permanently. For this reason some manufacturers cover the tip of the degausser's pole piece with a thin layer of plastic or rubber. If your degausser does not have this protective feature on it, it is a good idea to paint over the tip with a plastic material such as GC's Plastic Cote (see Fig. 10-4). When using the degausser, bring it close to the head and slowly back away to a distance of approximately five feet or more and then turn it off. This will leave the head demagnetized and ready to use. This procedure should be repeated for each head in the machine. If a machine is used daily, it is a good idea to clean and degauss the heads at least

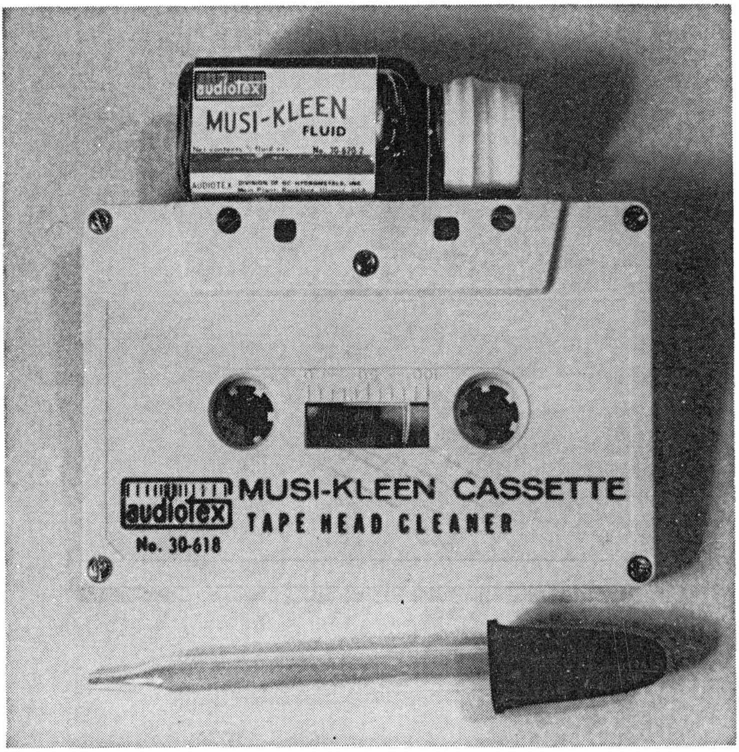

Fig. 10-2. A renewable, chemically treated head-cleaning cassette tape, this Audiotex item gets to head surfaces where swabs can't reach.

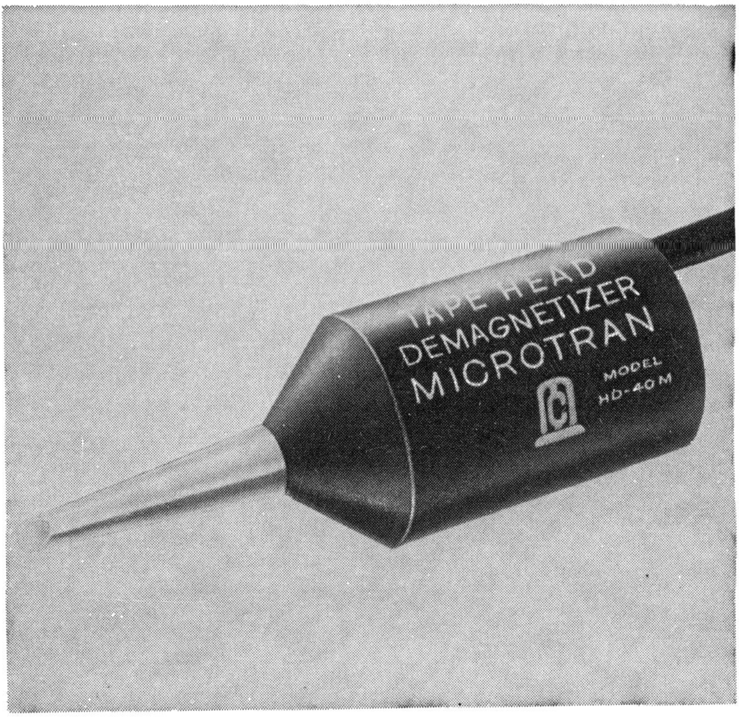

Fig. 10-3. Typical head degausser works on the ac line, but it is inconvenient in automobiles.

once a week. By way of comparison, professional users and duplicating laboratories clean tape heads after each reel of tape passes through the machine, and the heads are demagnetized at least daily.

Demagnetizing the heads in an automotive cassette machine is equally important but raises some special problems. Since the unit is usually not easily removed from the car, a special type of degausser must be used. All degaussers rely on alternating current to produce the fluctuating magnetic field needed to demagnetize heads. Since the voltage available in a car is dc, automotive degaussers must contain a small multivibrator circuit to produce the ac necessary at the pole piece. The same cautions should be observed here, although it's very often difficult to reach into the maw of an automotive unit and find the head exactly. For this reason a demagnetizing tape is sometimes called for in automotive applications.

BATTERY FAILURE

Certainly the most important single culprit in the failure of battery-operated tape recorders is the battery supply itself. Most portable cassette units provide some method of testing the batteries, such as a special section of the VU meter which doubles as a battery checker. Unfortunately, too many people try to make do with batteries that are somewhat weak, and this can result in a bad recording in which voices sound like Mickey Mouse during playback at normal speed. Another problem is that machines with weak batteries may not have enough pull to operate some types of cassettes, especially those with poor lubrication or a tendency to jam.

The moral of this story is: Check the batteries ahead of everything else when you encounter difficulties with a cassette tape recorder. Also, check for corrosion, since this can sneak up on a tape recorder battery supply very rapidly. The main culprit in this scenario is the alkaline battery, which tends to

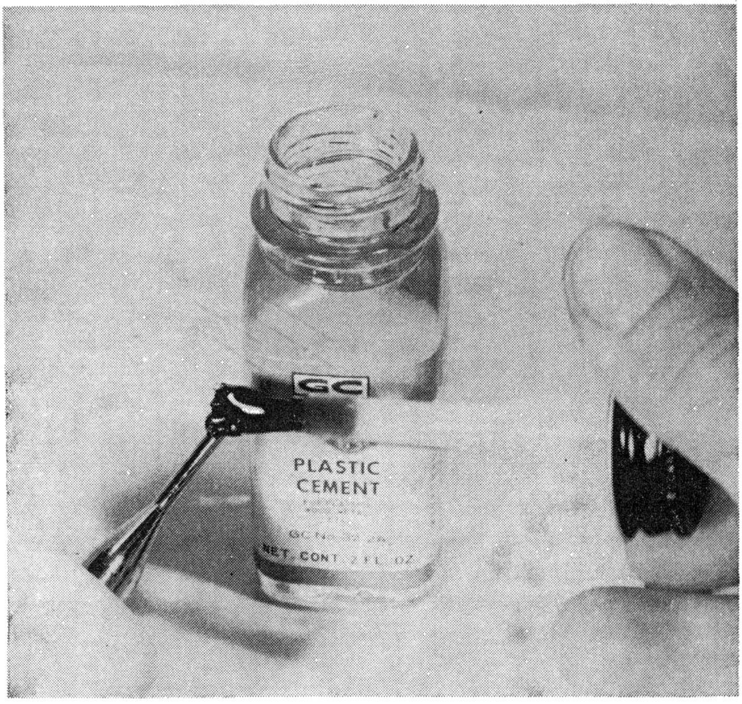

Fig. 10-4. To protect the heads from possible damage by physical contact with the degausser, coat the pole piece(s) with plastic.

secrete a very innocuous-looking liquid, even when perfectly sealed. This material turns into a flaky white powder, and while it generally will not corrode the terminals in the recorder, it does prevent good contact between the batteries and the brass contact strips. If you're using alkaline batteries, it's a good idea to remove them and sandpaper both ends of each battery lightly every time you use the tape recorder. If this sounds like a lot of trouble, you're right, and a good solution is simply to use good old-fashioned carbon-zinc or nickel-cadmium rechargeable batteries.

Another difficulty with batteries can occur in recorders that use clip-on or snap-on battery clips. Very often these become bent, loose, or otherwise worn in such a way that they do not make good battery contact. Always check such clips, including the bronze spring-metal type contacts to make certain that they are making adequate contact with the battery terminals.

CONNECTING CABLES

A frequently overlooked trouble spot is the connecting cable. Various types of cables are used in stereo systems at home, in the car, and for portable recorders. Most common types use an RCA pin plug (phono plug), or a 3½ mm phone plug, or a 2½ mm phone plug. These various types are shown in Fig. 10-5. Very often, interconnections must be made and jumper cables of an appropriate type must be made up or purchased.

It's usually best to use a molded, factory-assembled connecting cable, since these provide the best resistance to breakage and shorts. A typical cable assembly of this type is shown in Fig. 10-6. A simple check, if you suspect the cable of being intermittent, open, or shorted, is to use a VOM to check both for continuity and shorts. It's always best to flex the cable at the end near the plug while making these measurements; any meter fluctuations or wild jumps in the needle while flexing indicate an intermittent condition. In all such situations, it's best not to try to repair the cable, but to purchase an exact replacement.

Much routine servicing and maintenance simply calls for common sense in looking for the obvious. Of course, there are always some unobvious problems such as the infrequent tape

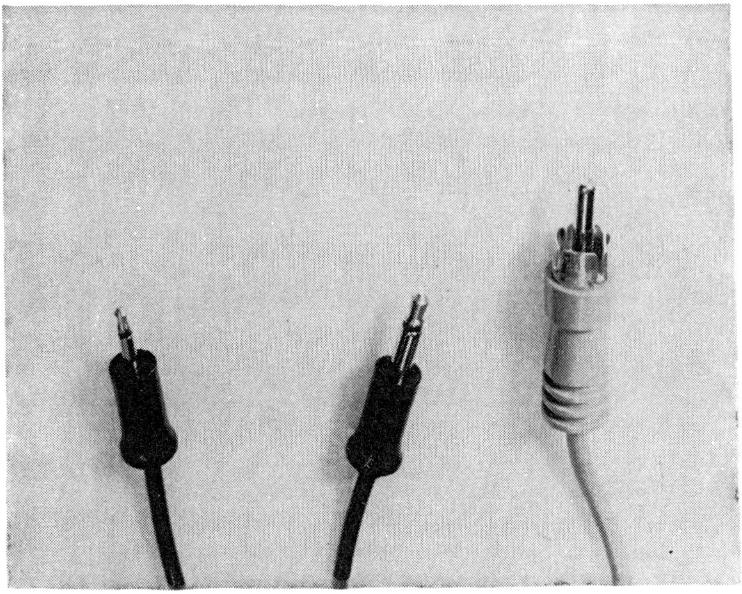

Fig. 10-5. Three common plugs: RCA phono type pin plug (left); 3½ mm phone plug (center); 2½ mm phone plug (right).

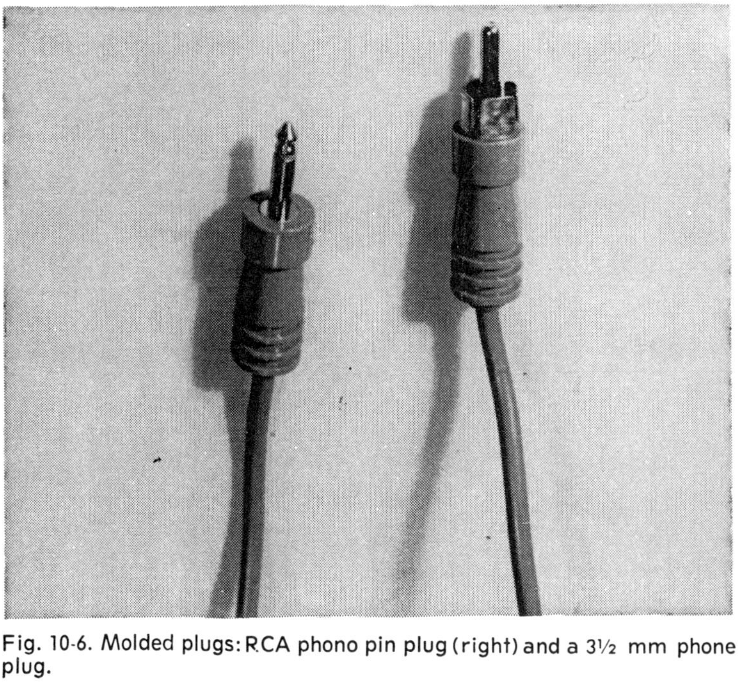

Fig. 10-6. Molded plugs: RCA phono pin plug (right) and a 3½ mm phone plug.

cassette that has the tape wound in backwards! You will also find other bad tapes from time to time which simply will not record for no apparent reason. If you do have a problem recording on a particular cassette and suspect that the tape itself might be at fault, the test for this is a very simple one—just try another tape. In a few cases, the tape formulation is badly matched to the bias and drive circuits of the recorder, and, therefore, will not record well no matter what you do.

If you are in a doubtful situation, check with the manufacturer of the instrument to see which type of tape will work best. Or if the manufacturer can't speak English, just try a few different types of tape formulations until you hit on one that works well.

SERVICING CASSETTE RECORDERS

Chapter 11

Servicing a cassette recorder is not too much different from working with open-reel type units, except that everything is on a much smaller scale. Most cassette recorders also have protective covers which very often impede access to the heads, capstan, and pinch roller, and other important guide mechanisms. Because of this, it's very often necessary to partly disassemble a cassette recorder even for the most perfunctory repairs.

This disassembly means removing the outer case, which sometimes is easier said than done. Every cassette manufacturer seems to have conspired in some special way to make the outer shell difficult to remove. If you don't have the individual service manual for the cassette recorder that you're handling, you may be in trouble. However, certain standard procedures are possible, and with a little bit of patience and forbearance, you'll probably be able to get the outside cover off without mishap. Covers are usually held in place by easily identifiable Phillips-head screws; there may be as few as four of them, or there may be a dozen or more. If the latter is the case, just grit your teeth and take a tranquilizer, or whatever else calms you down, and get to work.

TEST EQUIPMENT NEEDED

Generally, the test equipment you'll need is pretty much the same as for a standard tape recorder, depending on the extent of the repairs you expect to make. For electrical measurements and repairs, you'll need a VTVM, a frequency counter, an oscilloscope, an audio signal generator, and one or more test tapes, along with an appropriate blank cassette tape.

Fig. 11-1. Drive belts for capstan flywheel (large pulley at right) and for rewind (white plastic pulley) in a Pioneer stereo cassette deck.

When checking the equipment for mechanical problems, the first thing to look at is the drive belt. This belt (see Fig. 11-1) is subject to the usual aging and wear encountered with any other rubber component used in tape recorders and record players. If you determine that this belt is even slightly suspect, replace it instantly—especially if the machine is more than a couple of years old.

This belt may look like little more than a thick black rubber band, but don't be misled by its apparent simplicity. In some machines, square cross-section belts are used with differentiated sides. This is especially true of certain quality tape decks where two of the belt's surfaces are dull and two are shiny. Generally, the manufacturers recommend that the shiny surface be in contact with the drive motor pulley and the capstan pulley. They caution that if the wrong surface of the belt touches these pulleys, some slippage will result, causing more wow and flutter.

Wow and flutter are bad enough in most cassette recorders when they're working properly; anything you can do to improve this particular performance area can't help but benefit the overall system performance.

ELECTRONICS TESTING

If the equipment has serious electrical problems, then electronic procedures are in order. A basic test setup for checking a recording system is shown in Fig. 11-2. This basic setup can be used in one of several ways. One method is to record a 1000 Hz signal with the microphone or line-level volume control set to maximum and then play the tape back through a 1000 Hz bandpass filter. This will provide an excellent measurement of signal-to-noise ratio and just how much of a problem noise is in the tape recorder system. This same method can be used for checking the effectiveness of built-in Dolby circuitry by first measuring the noise level with the Dolby circuit off and then with the Dolby circuit turned on.

Dolby Testing

The block diagram of a typical machine using Dolby circuitry is shown in Fig. 11-3. Basically, the non-Dolby machine or the same machine with the Dolby circuit turned off are identical; just omit the Dolby squares from the illustration to visualize this situation. Dolby adjustments can be a little bit touchy and these should not be attempted without consulting both the owner's manual and the manufacturer's service manual. A Dolby circuit can be a relatively complicated affair, but it's not all that hard to handle; just follow the specific recommendations in the manual. A more comprehensive

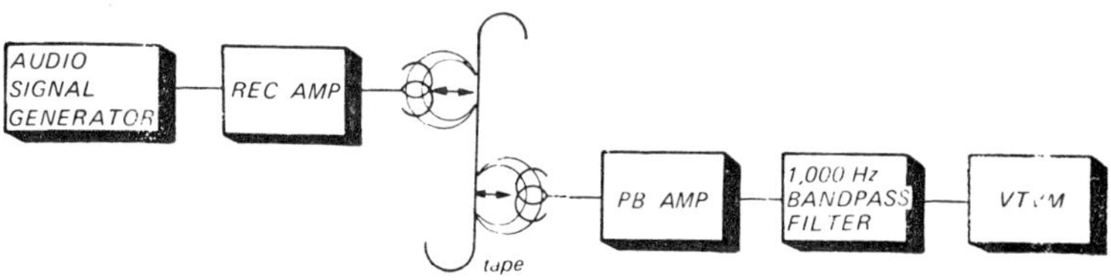

Fig. 11-2. Basic instrument setup for testing record and playback functions of any cassette recorder: The audio signal generator feeds a tone to the record amplifier and record head. The signal is played back from the tape through the playback head, amplifier and then through a 1000 Hz bandpass filter to a VTVM. The meter signal level indicates correct or incorrect alignment of the heads. (Courtesy Sansui.)

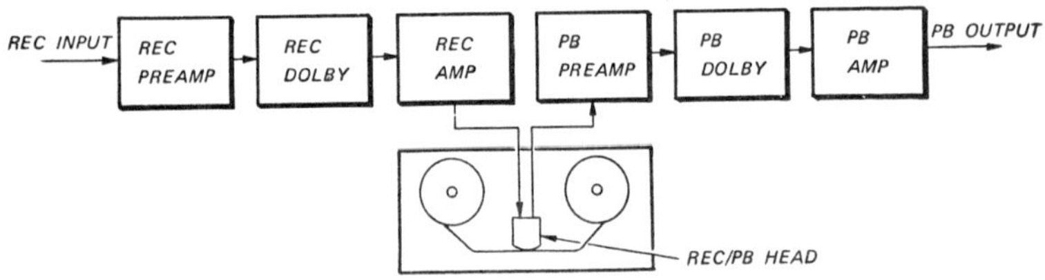

Fig. 11-3. Block diagram of a typical cassette stereo deck with switchable Dolby circuitry. (Courtesy Sansui.)

block diagram of a complete stereo deck with Dolby and motor servomechanism is shown in Fig. 11-4.

Specialized circuits seem to abound in today's sophisticated cassette recorders. The last one you want to cross and probably the one you should touch the least when servicing is the Dolby. Generally, this is to be considered as a "black box" with no adjustments to be made. However, there are a couple of touchup controls that occasionally might require just a light resetting. These adjustments are described in the owner's manual, since they are user adjustable. A typical Dolby circuit board used in a home stereo cassette deck is shown in Fig. 11-5.

Automatic Shutoff

You will find many other highly specialized circuits in today's sophisticated cassette machines. An example of this is the automatic shutoff circuit shown in Fig. 11-6. This particular circuit, while not that unusual, has one big difference from most other automatic shutoffs: Most shutoffs simply disengage the drive clutch and flash an automatic warning light of some kind. This particular shutoff really does what the name implies—it kicks the play button or fast-wind button back up into the off position and shuts the machine down, except for the power switch.

Bias Switching and Other Adjustments

Another very specialized circuit is the relatively complex switching needed to change from one bias setting to another. This will only be encountered on high-quality decks which

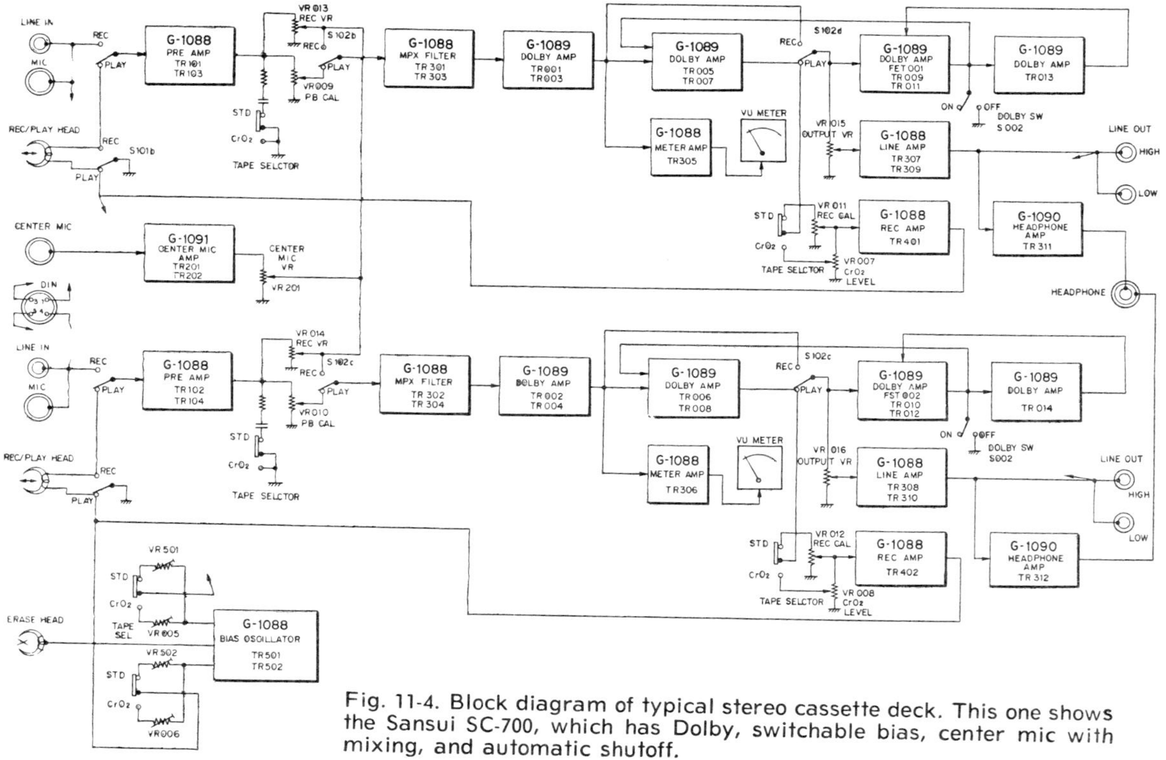

Fig. 11-4. Block diagram of typical stereo cassette deck. This one shows the Sansui SC-700, which has Dolby, switchable bias, center mic with mixing, and automatic shutoff.

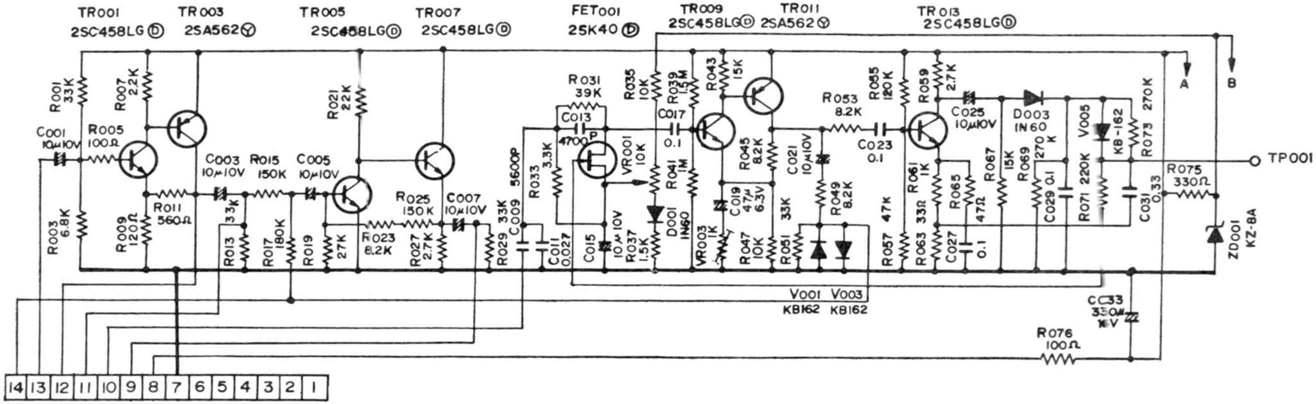

Fig. 11-5. Dolby circuit board. It is fairly identical to such circuits in all cassette decks that offer the Dolby noise-reduction feature. (Courtesy Sansui.)

provide switchable bias from that required for conventional ferric oxide to that needed for chromium dioxide or premium-grade ferric oxide tapes. The actual circuitry involved consists entirely of passive components, but these are permanently adjusted at the factory for the appropriate bias-level settings. Sometimes these adjustments can get out of whack, or the machine's requirements may change, in which case some readjustment may be needed. If you plan to reset the bias for any reason, be sure to use to use the appropriate kind of tape cassette for which the bias is to be recalibrated. A diagram of a typical bias switching circuit is shown in Fig. 11-7.

DRIVE MECHANISMS

Once the electrical problems are out of the way, the mechanical-drive system of the machine is next. Fortunately, the mechanics of the typical cassette machine are not that complicated (see Fig. 11-8). The underside of even the most sophisticated cassette machine is relatively simple when compared to a typical record changer, and you know what a headache those can be!

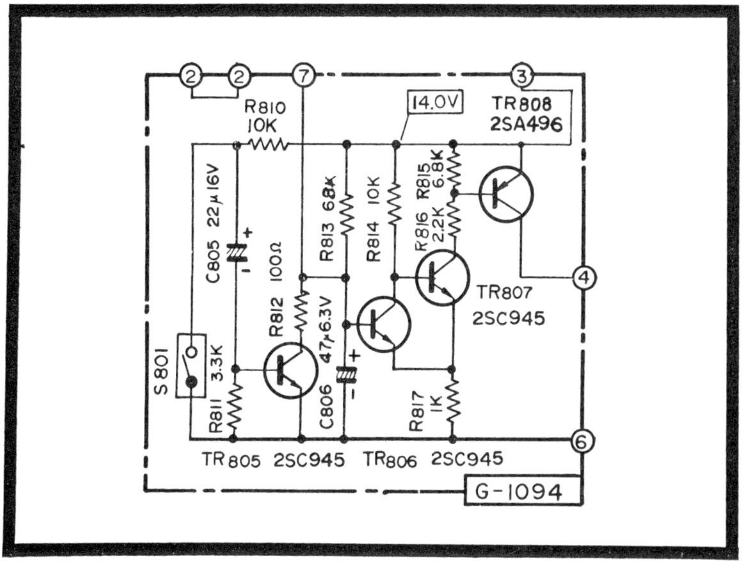

Fig. 11-6. Automatic shutoff circuit used in the Sansui SC-700. Similar types of circuits may operate flashing warning lamps, automatic reversing, and shutoff.

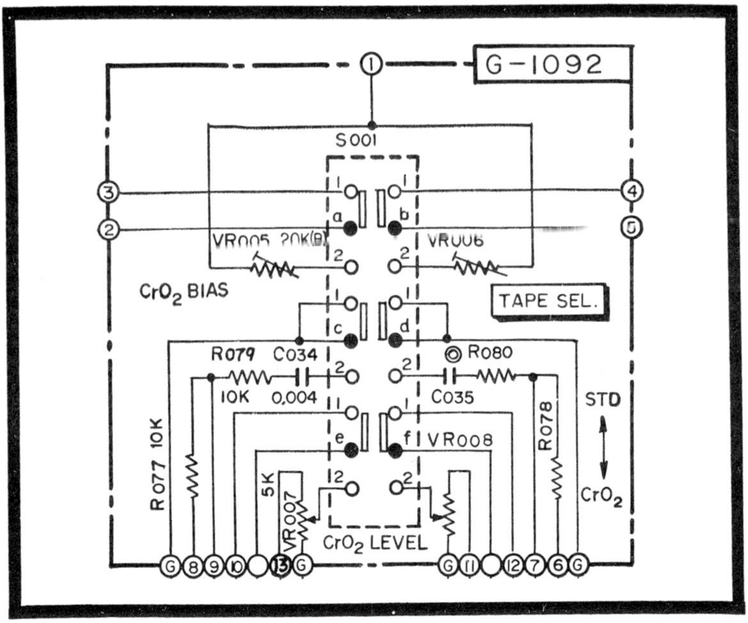

Fig. 11-7. Bias switching circuit for ferric oxide and chromium dioxide tapes. This circuit board appears in the Sansui SC-700, and is typical of most other such circuits in stereo cassette decks.

Yet, it's quite possible to run into more complicated drive mechanisms, such as the type shown in Fig. 11-9. Even here, the biggest problem is disassembling the mechanism, which is necessary even to install new belts. New rubber will clear up most of the problems in almost any cassette drive train.

The only other serious consideration with drive trains is proper cleaning and lubrication, which can be handled without any complex disassembly. Lubricants should be kept well away from any rubber and pulley surfaces that touch rubber drive elements. Lubricants should be high-viscosity types, such as Lubriplate or other similar paste lubricants. They should be used very sparingly, and just at the friction contact or bearing points in the drive train. Naturally, such a lubricant can be used more liberally in slides and other assemblies that are nowhere near the actual rubber drive elements.

Special Slip Clutches

Some machines use an unusual or special type of drive train quite unlike any other equipment on the market. One

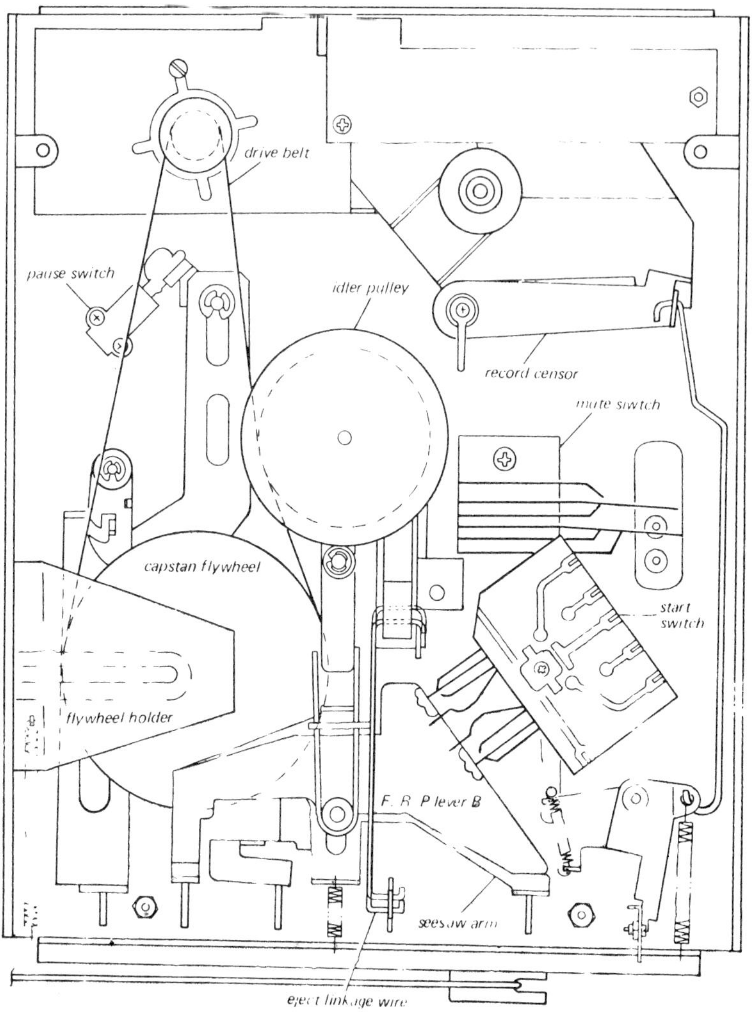

Fig. 11-8. Drawing of a typical under-deck mechanical assembly with cam-action switches and control functions. Most piano-key decks use some variation of this arrangement. (Courtesy Sansui.)

such special case is the unique clutch mechanism used by 3M in its tape recorders. This is detailed in Fig. 11-10. Also shown in Fig. 11-10 is a conventional drive train that uses belts and pulleys. Unlike these, the 3M drive uses entirely direct surface-to-surface contact between pulleys and flywheels.

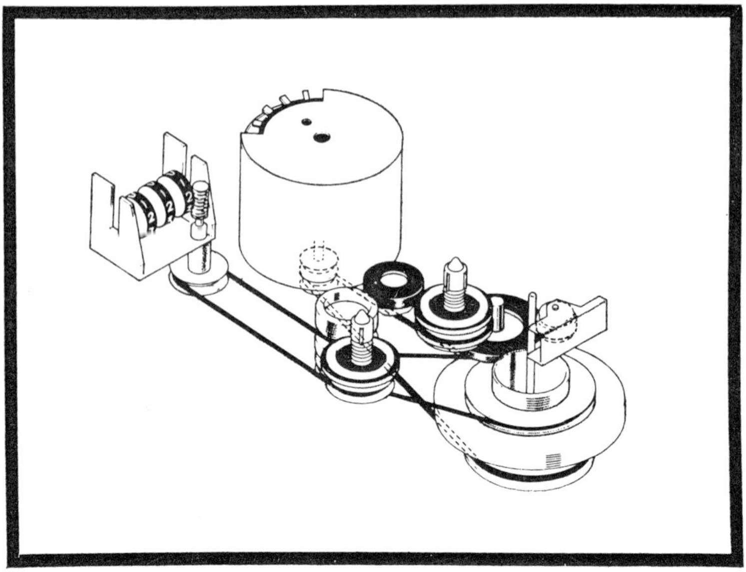

Fig. 11-9. Sketch of the pulley drive used in TEAC cassette decks.

There are pros and cons of such a system. Since there's no isolation of the motor from the drive train, some audio purists might claim that this transmits a certain amount of vibration to the cassette. Yet, this problem has been largely overcome by 3M in this equipment, since it works well without any undue vibration problems.

Finally, just before you put the cover back on the machine, check the head alignment and performance with an alignment test tape. The adjustment screws are on both sides of the head as shown in Fig. 11-11, and only a tiny touch with a screwdriver is needed in most cases. After adjusting the screws to your satisfaction, be sure to dope them with a sealing paint or bright-colored nail polish.

If you have the urge to develop a total feel for cassette equipment mechanisms and how they work before going ahead with any repairs, you might try building a kit. The only one on the market right now is made by Heath and it is probably typical of basic cassette drive mechanisms. You can't help but learn a lot about how these machines work when you put together a unit such as this.

Other than straightforward cleaning, lubricating, and adjusting, any specialized repairs that may be needed in

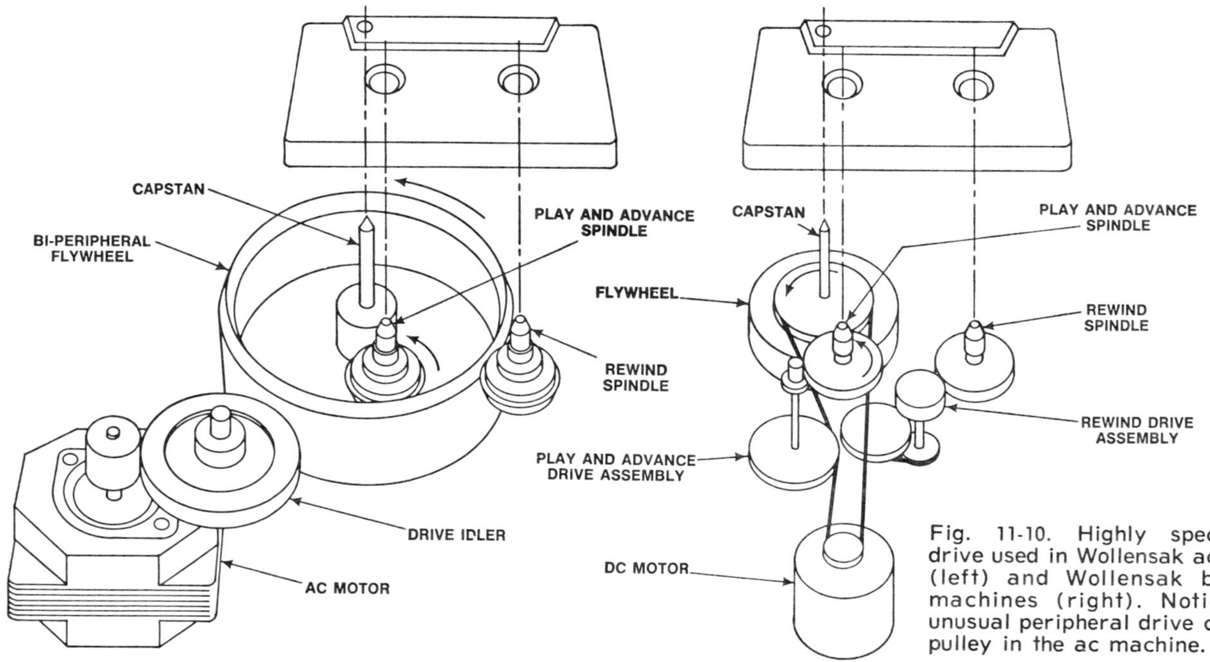

Fig. 11-10. Highly specialized drive used in Wollensak ac decks (left) and Wollensak battery machines (right). Notice the unusual peripheral drive capstan pulley in the ac machine.

Fig. 11-11. Where it's all at—the head and drive assembly of the typical cassette recorder. You'll seldom see all the parts this clean on any machine that comes in for servicing. (Courtesy Pioneer.)

a cassette recorder may definitely require the use of a service manual. We've mentioned service manuals before in this chapter and will say it again: Get a copy of the manual before attempting any extensive repairs, and certainly before doing any critical electrical adjustments on equipment with highly specialized circuits.

OFFICE AND BUSINESS MACHINES

Chapter 12

One of the earliest applications for the Edison phonograph was as a dictating machine. In fact, Edison Voicewriter cylinders persisted in some offices until only a few short years ago. But there are many better and more convenient systems for office dictation and transcribing today; the most convenient of all of these media has turned out to be magnetic tape.

Magnetic recording has been used for office systems for a number of years; Poulson's Magnetophone was the predecessor of today's tape systems. Now, there are dozens of magnetic tape systems available in various sizes, types, and formats. The cassette is a late bloomer in this field because of one very severe mechanical limitation: until fairly recently, there was no foolproof method of backing up the cassette by footpedal for repeating material. Certainly, cassettes could be reversed by pushing the rewind button, but there simply had been no way of doing this from a footpedal.

REPEAT TAPE

One of the first machines to provide a method of repeat playback was the Crownscriber (see Fig. 12-1). This machine, which is a very convenient system and very low-priced (about $130), uses a closed loop of tape within the machine which constantly records what is being listened to by the typist. This loop, when triggered by the remote-control footpedal, will play back the last seven seconds of recorded material. It's an ingenious system, but some typists have complained that the seven seconds is not long enough, although this complaint is not a common one.

Many businessmen have found the cassette so convenient a format thay they started using cassette dictation before appropriate transcribing equipment was available. Very

Fig. 12-1. The Crownscriber shown here was the first cassette office transcriber that offered repeat play. The secret: an endless loop of tape inside, constantly recording the material being played back for seven seconds' worth of repeat playback time.

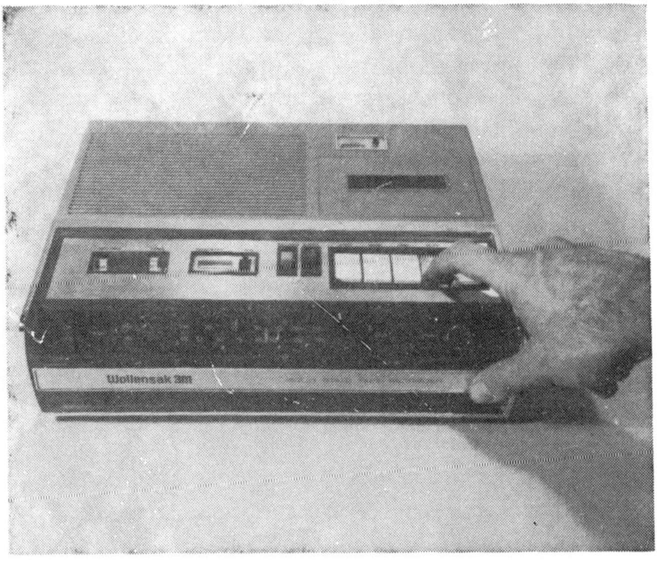

Fig. 12-2. A Wollensak entertainment-type machine is typical of units that can do double-duty as a dictating unit and playback transcriber. Its piano key controls offer instant backup and play with little or no time lost by typist.

often, a typist would be given an ac-dc machine with piano-key controls such as the one shown in Fig. 12-2. With a machine of this type, it's possible to reverse the tape and go back almost any distance. And with a little bit of practice, the pushbutton controls become a very convenient way of doing this.

POCKET PORTABLES

One of the most ingratiating aspects of using the cassette in business is the fact that almost anything can be used for making the original recording—including the new breed of pocket-size cassette recorders (see Fig. 12-3). Machines of this

Fig. 12-3. A pocket-size cassette machine, this Wollensak 400 has built-in mic and automatic recording level control for on-the-go dictation.

type are so incredibly convenient and lightweight that they are beginning to blossom in the most unlikely places.

An ultra-portable such as the Wollensak 401 series does have certain drawbacks. The biggest problem with a machine like this is the fact that it eats batteries alive. This is understandable. since it uses a tiny cluster of four penlight batteries, and no matter how carefully the motor of such a recorder is designed, it is going to use these batteries up very quickly.

In view of this, the Wollensak and many recorders like it are packaged with an ac adapter which can be used whenever the machine is near an ac outlet. Also included is an automobile cigarette lighter adapter. These adapters cover the power needs so well that the only time the batteries really need to be used is when the convenience of true portability is called for.

A machine of this type is lightweight and convenient and can be carried anywhere in a briefcase or in a raincoat pocket, although it might make an unsightly bulge in a jacket pocket. It's so light and so incredibly easy to use that more and more businessmen are turning to this type of recorder.

Typically, a pocket recorder of this type has a built-in microphone, so that even the dangling microphone cord is eliminated. An accessory microphone with a remote-control switch is provided, mainly for the convenience of using the remote switch.

Still another entry in the miniaturization race is the tiny Norelco mini-cassette shown in Fig. 12-4. This machine, called the "Pocket Memo," uses a nonstandard cassette which is so tiny that it only holds about 15 minutes worth of dictation or note-taking on each side. The machine is truly a pocket recorder, but it is not compatible with other systems; very often the cassette must be dubbed onto a standard-size cassette before it can be transcribed. This unit is especially good for taking notes for later reminders. It really is not intended to be used as a dictating medium.

FAST TRANSCRIBER

One of the most important contributions to cassette office equipment is the Wollensak audio-visual series. The standard A-V recorder equipped with a pulsed footpedal (see Fig. 12-5)

Fig. 12-4. Norelco "Pocket Memo" cassette recorder uses a nonstandard cassette which holds just 15 minutes worth of tape per side. But the machine is ultrasmall for convenience note-taking.

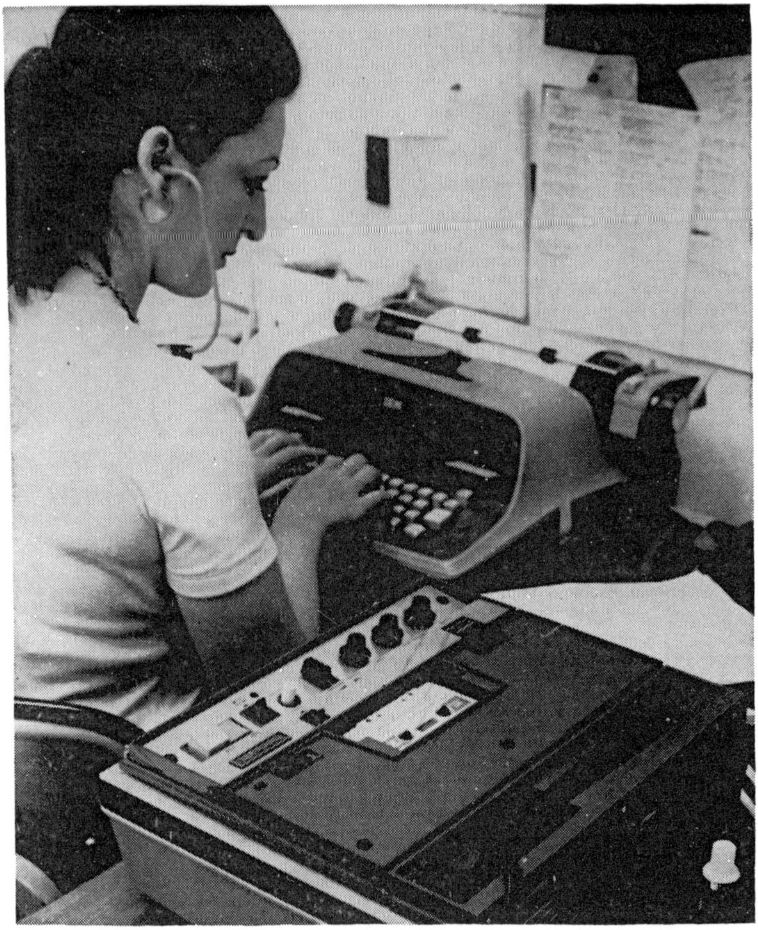

Fig. 12-5. Wollensak audio-visual type machine is ideal for office use. Solenoid-operated head and pinch roller assembly retracts when the unit is not playing, and fast-forward and rewind functions can be controlled by footpedal.

represents a major advance in equipment designed for office use. Though not intended originally as an office machine, this unit does the job so well that one wonders why someone hasn't been able to come up with a similar design sooner.

The ingenious part of this machine is the fact that the entire head and capstan assembly is solenoid-controlled. When the solenoid is operated, the head and capstan assembly pushes forward to make contact with the tape. As soon as the playback footpedal is released, the entire assembly retracts, pulling away from the tape. Either of two fast-wind levers on the footpedal can be pressed for incredibly fast rewind or fast forward. This is the only system on the market that permits this type of fast transport by remote control (Fig. 12-6).

The one unfortunate feature of this machine is its bulk. It is huge, and some secretaries object to having the monster on their desks. But hidden under that bulky exterior is one of the most rugged cassette transports on the market, and the machine is well worth the somewhat higher-than-average price that it commands.

CAR CASSETTE UNITS

One possibly often-overlooked business machine is the automotive stereo cassette player-recorder. While this is really intended as an entertainment machine, the fact that it has a recording function makes it ideal for the businessman on the go who wants to dictate in the car. A typical machine of

Fig. 12-6. Accessory jacks of the Wollensak A-V machine are designed for a special 4-prong plug for the pulsed foot-control unit.

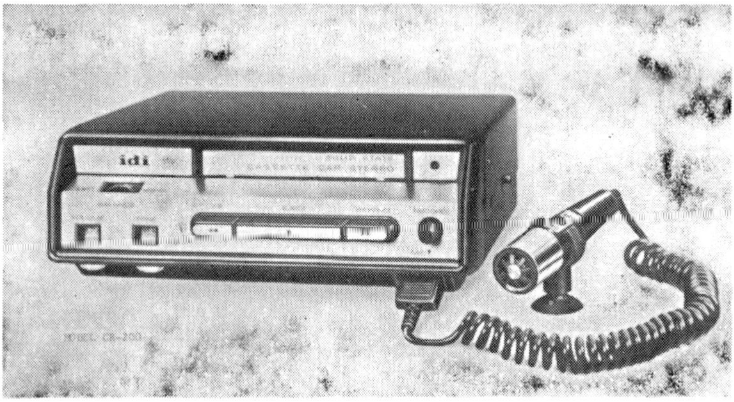

Fig. 12-7. The automotive stereo player-mono recorder is gaining popularity as a business machine. (Courtesy Inland Dynatronics.)

this type is shown in Fig. 12-7. The obvious advantage here is that it prevents having a lot of loose equipment hanging around the car. There's no need to plug cables into cigarette lighter sockets, and the microphone stows away very neatly when you're done with your work—all without having to shift materials and equipment around with your free hand on the front seat.

BUSINESS ADVANTAGES

The whole idea behind using cassettes in business is the fact that they are convenient. Convenience is something that has been lacking in dictating machines and business equipment that it still is a little bit startling to see what really can be done with designs of this type.

This is an era of miniaturization, and the cassette is one of the beneficiaries of many new design trends today. The integrated circuit has helped to slim down not only cassette recorders but electronic computers and calculators, which are now made in pocket size models, and just about everything else that only a short time ago took up plenty of extra space.

One other important factor with business equipment is the cassette itself. Naturally, cassettes are much cheaper than special tapes that must be purchased for each specific type of magnetic office dictating system. In fact, they are so cheap that the tendency is to look for very inexpensive, bargain-basement items, thinking that this is going to save a lot of money. At some point, the law of diminishing returns sets in and the bargains turn out not to be bargains at all, but sources of difficulty and annoyance that cause trouble.

It's a good idea to invest just a little extra in good-quality cassettes with foolproof and jamproof internal construction. The few pennies extra spent on each cassettte will reap vast benefits in time and aggravation that you won't have.

LIVE RECORDING AND DUBBING

Chapter 13

Initially, the cassette tape recorder was intended to be a portable machine for on-location recording of voice, dictation, and other applications that call for an extremely high degree of portability. The very first such machine—the Norelco 150 (see Fig. 13-1), fulfilled this concept very nicely. It had a self-contained battery supply and it was relatively light in weight as compared to other portables of its day. This portability was underscored by the fact that all such early machines had a shoulder strap and carrying case, and was the very first of a class of truly portable recorders to hit the market.

These early machines left a great deal to be desired as far as fidelity was concerned. They came equipped with dynamic microphones, simply because the impedance characteristics of the dynamic mike tend to match those of the transistor circuitry used in these recorders. Certainly, basic microphone techniques are not of paramount importance with portable machines, especially when they're used to record speeches and interviews or for notetaking, dictation, etc. All you really need is to get a clear and intelligible signal on the tape. You can place the microphone almost anywhere that sound pickup is loud enough to make the needle move when recording.

But in situations where you have poor direct pickup, such as a large hall where there is some distance between the recorder and the speaker, or when trying to interview someone at opposite ends of a long conference table, certain recording problems come up.

AUTOMATIC LEVEL CONTROL

The average portable cassette recorder, provided it's not one of the really inexpensive ones, is equipped with an automatic level control which takes just about all the sweat out of recording in different situations. What the automatic level control (alc) does is to make sure that a sufficient recording level is always received by the tape—within the limits of the microphone's pickup. Sudden loud noises will cause the recording level to drop drastically and it will come back up during periods of silence or lower-volume speech. In effect, this marvelous circuit "rides the gain," and there's no need to watch a recording level meter. A typical alc circuit is shown in Fig. 13-2.

Microphone placement in a conference room, or in any recording situation, depends largely on whether or not the recorder is equipped with an alc circuit. If it does have such a circuit, then microphone placement isn't nearly as critical as it is when manual monitoring of the VU meter is necessary. In the latter case, it's important to try to place the microphone midway between the two people in a typical conference or interview situation. With an alc circuit in use, the microphone

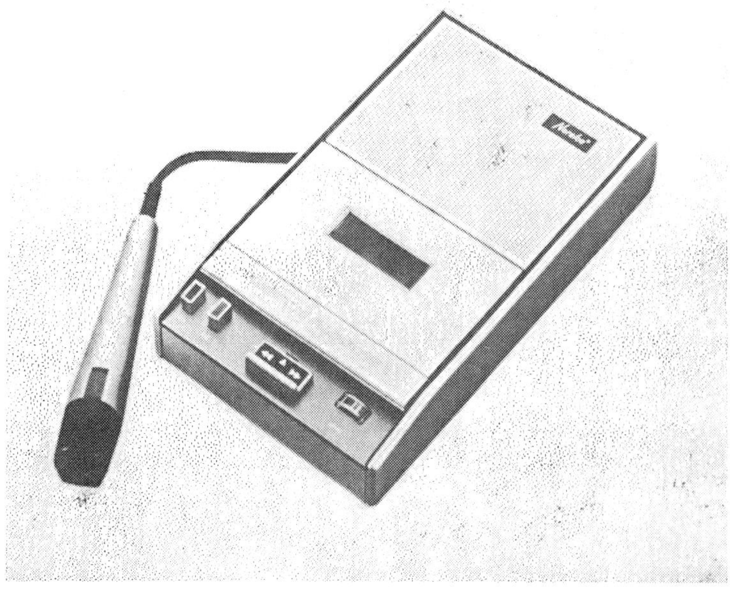

Fig. 13-1. Typical cassette recorder—the redesigned Norelco Carry-Corder and outboard microphone with remote switch.

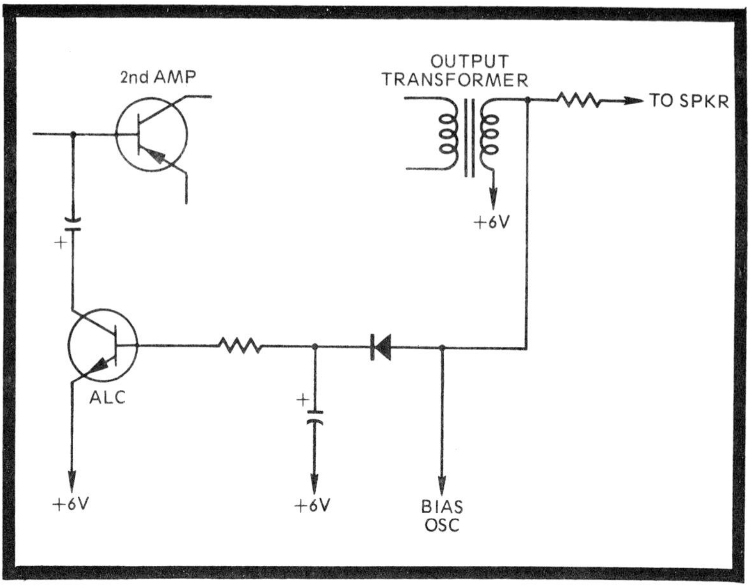

Fig. 13-2. An automatic level control circuit (alc) is basically a feedback/compression type, which prevents the recording signal from overloading the record head and tape.

placement is not as important, although the same general rule should be applied.

MICROPHONE DIRECTIONALITY

Most microphones supplied with cassette recorders are **omnidirectional**. This means that they will pick up sound equally well from any direction—even from behind. This is a normal characteristic of a dynamic microphone. However, a dynamic mic can be designed to almost any directional specification by adding special vents, openings for back pressure, or other special features. Generally, such a modified microphone—called a directional microphone—costs appreciably more than its omnidirectional counterpart.

Microphones for portable tape recorders contain a switch for turning the tape recorder motor on and off. The most common type, imported from Japan, uses a two-prong plug (see Fig. 13-3) at the end of the connecting cable. The larger of the two prongs or plugs is for the microphone itself; the smaller one operates the tape recorder motor through a closed-circuit jack of the type shown in Fig. 13-4. The

Fig. 13-3. The pencil-thin dynamic mic and remote control switch for the Wollensak pocket recorder is typical of the miniaturization and convenience found in today's cassette recorders.

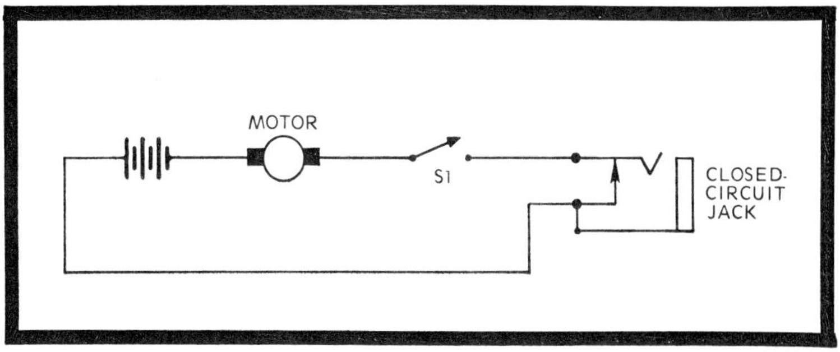

Fig. 13-4. The remote control switch operates the recorder motor through a series-connected closed-circuit jack.

microphone plug diameter is 3½ mm and the remote switch plug has a 2½ mm diameter. They are often referred to as miniature and subminiature phone plugs, respectively.

If the recorder is to be used with a footswitch for playback, the footswitch will be fitted with a 2½ mm plug and will go into the same jack that had been occupied by the smaller of the two plugs on the microphone cord. This type of footswitch simply provides an on-off action for the recorder; it does not provide special features such as backup or repeat play. Only special types of office transcribing machines (covered in Chapter 12) have this capability.

HOME STEREO MICROPHONES

There are several types of equipment packages in home stereo recorders. Complete units—those with built-in amplifiers and loudspeakers—generally come equipped with a pair of dynamic microphones. Cassette decks, on the other hand, usually do not have microphones in the factory package, even though a provision is made for connecting them.

The difference between the two types of units is obvious in another way. The recorder that is equipped with its own microphones will usually have 3½ mm (miniature) phone jacks while the deck will have standard-size phone jacks for standard (full-size) microphone plugs. The reason for this is because the deck will likely be used for more serious recording and, therefore, should be adaptable to higher quality microphones, which are usually equipped with larger plugs (see Fig. 13-5).

There is at least one stereo deck on the market that has provision for a third microphone (Fig. 13-6) for the so-called center channel and a separate recording level control for blending or positioning that microphone electronically in or between the two stereo channels.

POLAR PATTERNS

If the question comes up about buying microphones to use with your home cassette deck, try to stick with the dynamic type, since these are the most rugged and versatile. They are available in the omnidirectional type, which has the kind of pickup pattern shown in Fig. 13-7A. An alternative type of

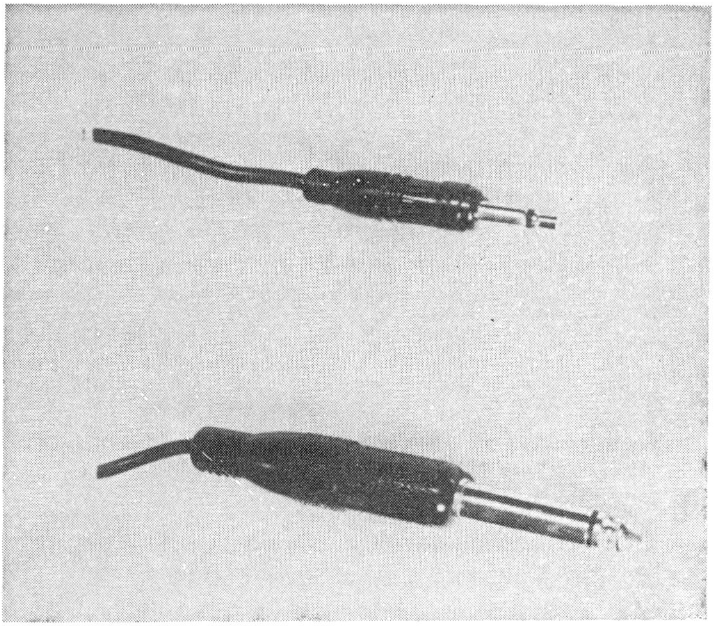

Fig. 13-5. Standard (PL-55) phone plug and miniature, 3½ mm plug are both used in today's cassette recorders. The larger of the two is usually used only with ac-operated stereo decks.

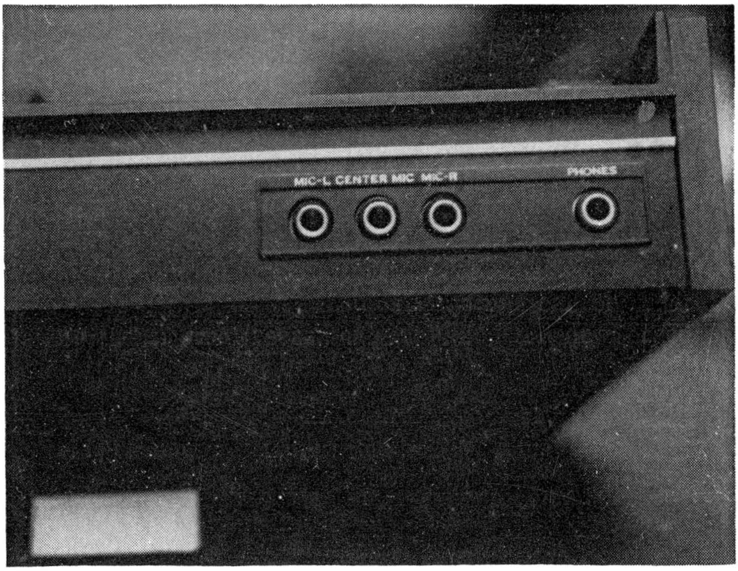

Fig. 13-6. Typical input-jack array on a Sansui SC-700 stereo deck—left, center, and right mic inputs, and headphone output.

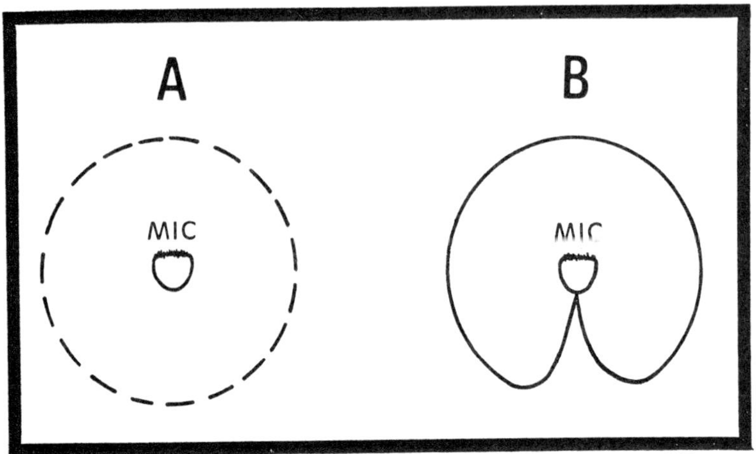

Fig. 13-7. Polar patterns for two types of dynamic mics: (A) omnidirectional and (B) cardioid.

pickup pattern is the **cardioid** (see Fig. 13-7B). The cardioid pattern gives the microphone a certain amount of directionality. It is completely dead to sounds coming from the rear, and its sensitivity gradually increases around the sides; directly in front, the microphone is most sensitive. A typical cardioid microphone is shown in Fig. 13-8.

CARDIOID MICROPHONES

In some situations, the cardioid may be called for. You may want to use a microphone in a noisy location, and this is where the cardioid's properties are very helpful. It also might have to do double duty as part of a public address system, and in such cases the cardioid is not merely helpful—it's mandatory.

Study all of the characteristics carefully on the mic specification sheet, and be sure to pick a unit that has a high enough output level to drive your recorder adequately. Some microphones produce a notoriously weak signal and often not even the spec sheet will let you in on this little secret

A new class of microphones has appeared very recently. These are the Electret condenser types. Some of the newest portable tape recorders coming from Japan have such microphones built into the tape recorder case itself. Such microphones are capable of far superior frequency response when used properly, and are very inexpensive as compared to conventional condenser microphones.

MICROPHONE FREQUENCY RESPONSE

Microphone frequency response is a very important factor when making your selection. Even though two microphones might look alike and even have similar price tags, the frequency response characteristic of each can vary widely.

A typical high-end response for an average dynamic microphone is about 10,000 to 12,000 Hz. By spending a few more bucks you can raise this upper limit to about 15,000 Hz, which might be a little too much frequency response for the equipment you're using—especially if you don't use a premium-grade tape. If directionality is not important,

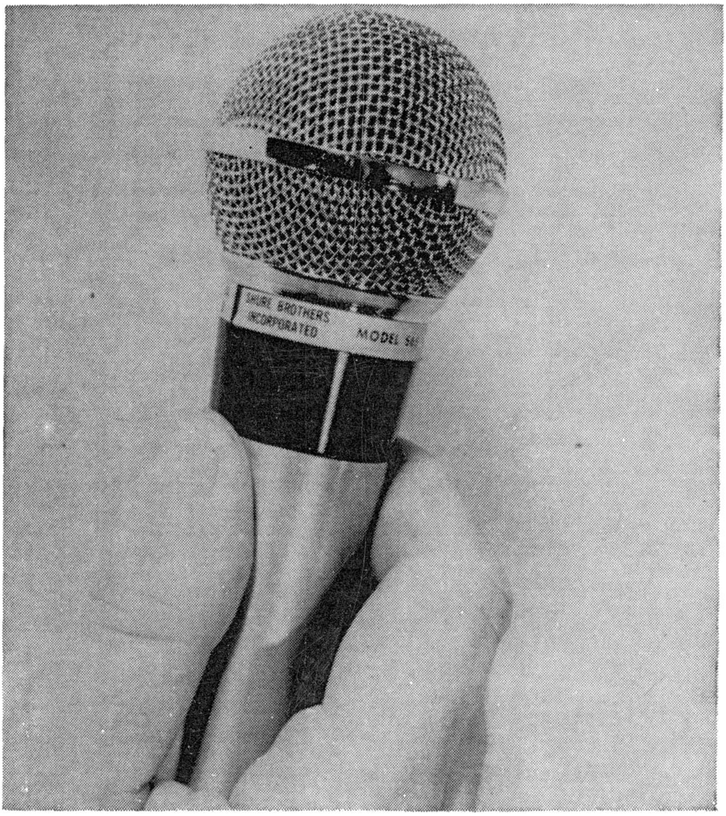

Fig. 13-8. A typical quality cardioid mic is this Shure 565.

remember that you can get more frequency response for the same money in an omnidirectional microphone than you can with a cardioid type.

DUBBING TAPES

Probably the most satisfying type of recording that you can do is to dub from other tapes or from records or the radio. This type of recording can produce excellent results because you're using professional equipment and techniques—the microphones, musicians, and studios that went into making the original recording. Given reasonable care and the right kind of equipment, there is no reason why your own recording can't be fully as good as the original, with the exception of a little bit of tape hiss.

The most common dubbing setup is shown in Fig. 13-9. Here, a direct connection is made to the tape output jacks on the rear of your stereo amplifier, which provide a signal output for each of the two stereo channels in your cassette recorder. You don't have to stick with just a deck to make these direct connections; a stereo recorder, possibly even a "portable" type, can also be connected directly to the tape output jacks. The only thing you have to do is be careful about possible feedback by shutting off the recorder's loudspeakers or adjusting the volume so it will not interfere with monitoring the source material.

When dubbing on a standard deck, it's a good idea to make a dry run from your program source with the controls set in the record position and the pause control lever depressed. This way, all of the record circuits are switched into the line, but the tape isn't going anywhere. Some equipment doesn't have this pause feature, but most stereo decks do, and it should be used to advantage.

With your controls set this way, watch your VU meters and adjust the recording level separately for both channels so you get approximately the same amount of swing on loud music passages. Naturally, both meters are not going to be exactly the same, since they represent two different channels of sound, but they should average about the same amount of swing.

OVERMODULATING AND TAPE HISS

It's also safe to go into the red area of the VU meter with some of the new high-coercivity premium tapes without causing distortion. This may go against the grain, especially if you've done any amount of home recording in the past. But try it; you'll see, it works and you'll end up with a cleaner, less noisy recording.

One of the first things you'll notice if you do record and tend to ride into the danger area of the VU meter with a premium tape is a reduction of tape hiss during playback. By recording at this higher level, you have, in effect, placed a higher signal level on the tape, and you will be able to turn down the playback level control when listening to your recording. As the playback volume decreases so does the noise level. This technique is recommended by at least one tape manufacturer (TDK) and it works admirably.

Some of the new breed of home recorders include Dolby noise reduction circuitry. This system further reduces tape hiss and will let your cassette recording approach reel-to-reel quality. (Dolby circuits are discussed in detail in Chapter 4.)

There are two schools of thought about the number of VU meters necessary on home equipment. Some rather expensive decks contain only one such meter and may or may not have a

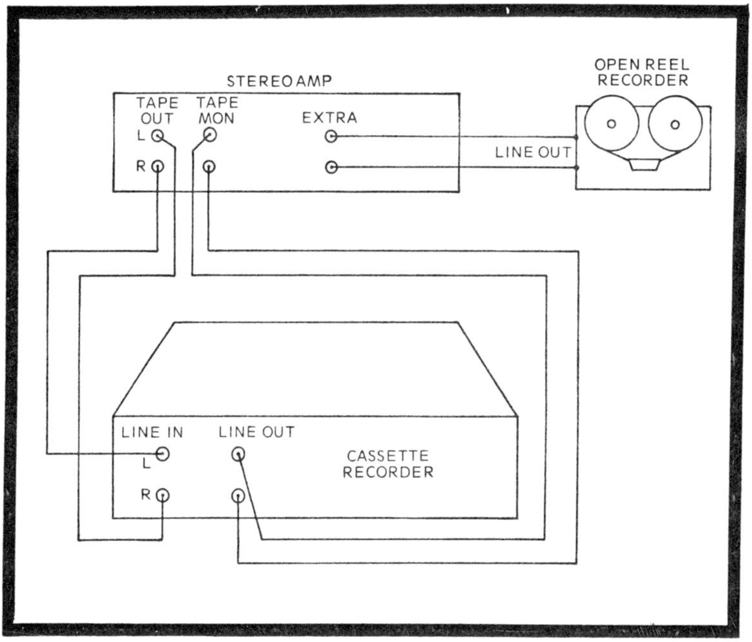

Fig. 13-9. Diagram for hooking up a cassette stereo deck for dubbing from an amplifier's tape output terminals.

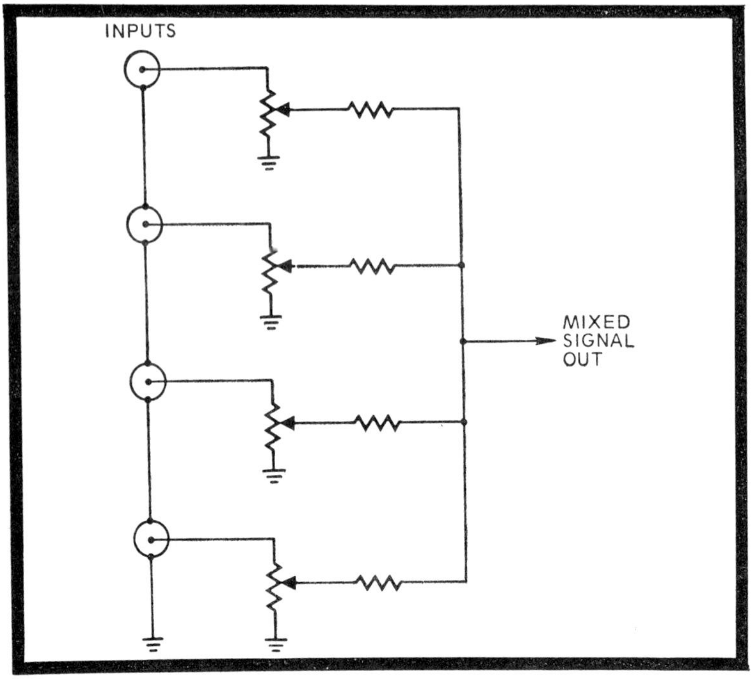

Fig. 3-10. The simplest type of mic mixer has a variable resistor in each input and produces a combined single-wire output.

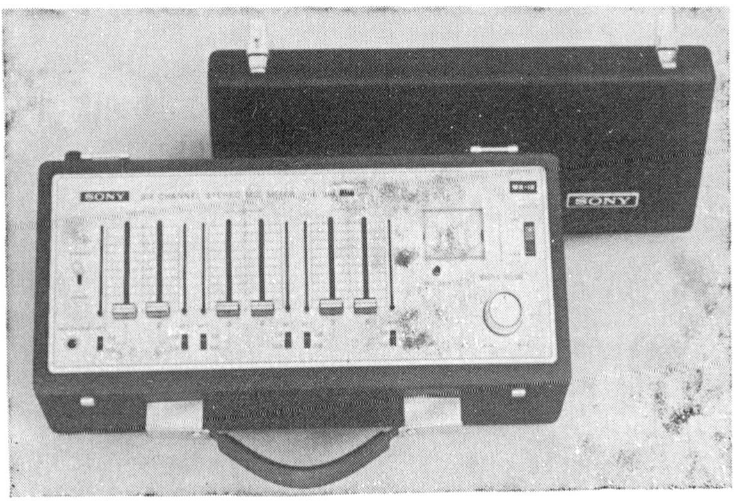

Fig. 13-11. This sophisticated powered mixer from Sony operates in either the stereo or mono mode.

meter selector switch. If there is no switch, this meter will generally show the highest swing of the two channels. A selector switch lets you sample the left or right channel or both. When set in the two-channel position, the meter reads the higher of the two.

Generally, a single meter is not a good idea and should be avoided if you're buying a deck for the first time. Yet, it will work, and it's possible to get good recordings, but you'll never know if both channels are working properly. One of them could have a bad connection and be giving you a very weak signal. With a single meter, you won't know about this until you play the tape back.

MIXERS

From time to time, you may want to mix together more than two stereo signal sources, such as recording with four or more microphones. This is also true when recording monophonically from several different microphone sources, such as with a panel discussion.

There are numerous mixers on the market which can handle this function quite well. Most of the less expensive units use a simple resistive network as shown in Fig. 13-10. Such **passive** mixers can do the job, but because they also decrease the available signal level, the noise level can increase on the tape—sometimes drastically.

To solve this noise problem, more sophisticated units, such as the one shown in Fig. 13-11, use active circuitry—a transistor amplifier section either before or after the resistive network. Ideally, each input should have a FET (field-effect transistor) amplifier to overcome this noise problem, but such a design is relatively costly, except in home-built units. Most mixers use, instead, a single transistor amplifier in each output. While most quality mixers are monophonic, it's a fairly simple matter to stack them in pairs and use them for stereo.

Chapter 14

RECORDING FROM THE TELEPHONE

Recording from the phone can be a source of pleasure, as well as an invaluable aid and timesaver in many business and personal activities. It's used by law-enforcement agencies and private detectives either armed with court orders or illegally; but it's done all the time, and legitimate uses are almost endless.

We've talked about recording from the phone in a rather off-handed way in other chapters without being very specific. Unfortunately, it's not always possible just to plug a recorder into a phone and to let the tape run.

INDUCTION PICKUPS

The most usual way to pick up the signal needed from the telephone is with an induction coil, attached to the telephone handset with a suction cup as shown in Fig. 14-1. The illustration shows the best possible location for one of the suction pickups. An alternate location is shown in Fig. 14-2. This spot has to be discovered by experimentation, since it relies on the magnetic field around the induction coil in the phone body, and this position can vary from one instrument to another.

The main reason for using the pickup on the side of the phone is that it eliminates the annoyance of the cord getting tangled with the handset cord and otherwise getting in the way during the phone conversation. The disadvantage: Not as strong a signal is developed at this point as in the handset, and the recording may not be good.

A fairly new method of attaching an induction pickup to the telephone handset is shown in Fig. 14-3. This slip-on arrangement fits on the telephone receiver like a glove and is not continuously falling off the way a conventional suction cup

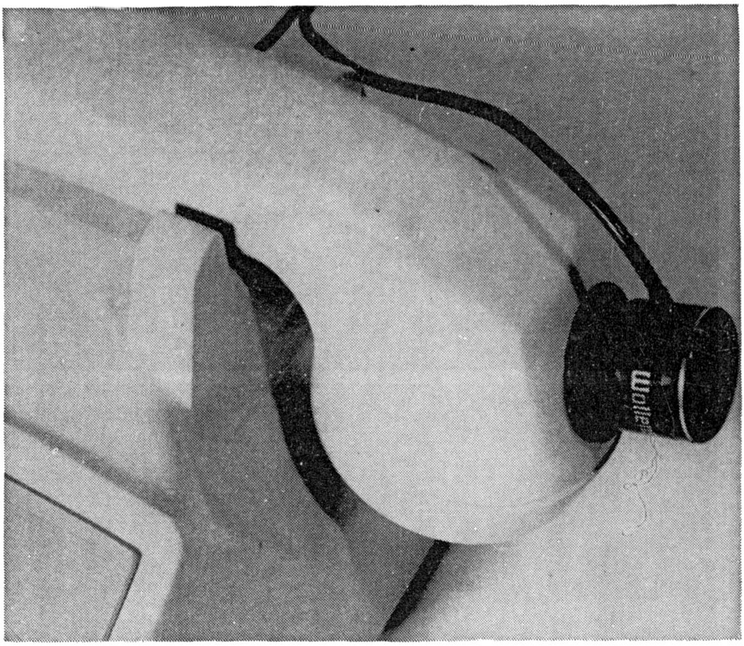

Fig. 14-1. A suction telephone pickup is usually placed on the receiver end of the handset.

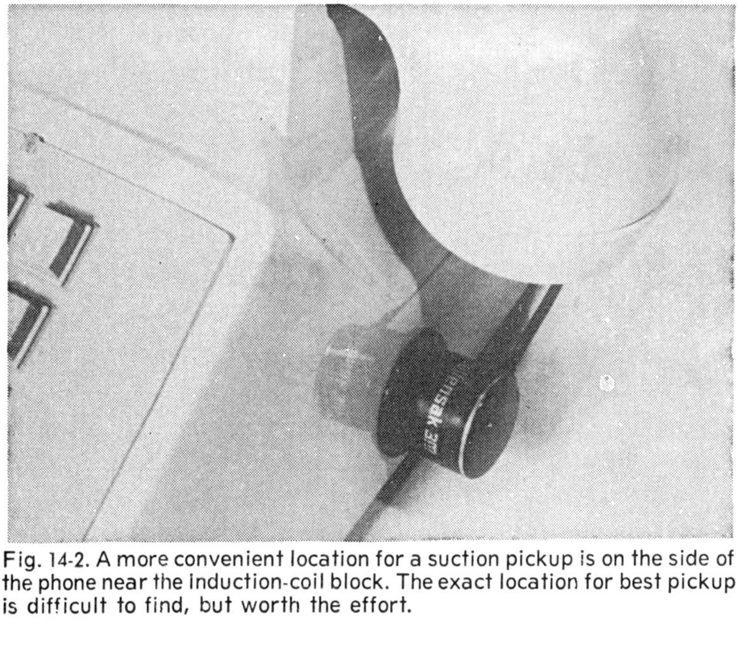

Fig. 14-2. A more convenient location for a suction pickup is on the side of the phone near the induction-coil block. The exact location for best pickup is difficult to find, but worth the effort.

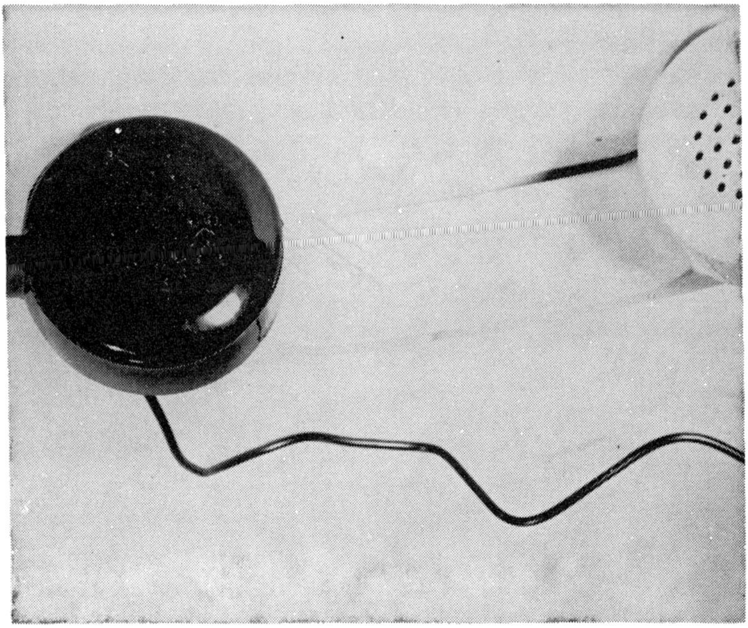

Fig. 14-3. The newest type induction pickup slips over the receiver end of the handset, but the dangling cord can still get in the way.

type of unit will do. A telephone pickup of this type can be left on the phone almost permanently and hardly ever gets in your way, except for that dangling cord which, of course, can get tangled up with the telephone cord as with any induction pickup.

An alternate type of magnetic pickup is a flat, Bakelite-enclosed device designed to fit underneath the telephone, directly under the induction coil block. While such a device is very convenient, its pickup is limited because of the physical distance from the block and the fact that it's difficult to position it exactly.

If you expect to do a lot of telephone recording, you might install one of these blocks semi-permanently. Break open the plastic case that encloses it; remove the flat coil and tape it directly to the telephone induction coil inside the telephone case with plastic electrical tape. This will provide a semi-permanent connection to your telephone, which will be extremely convenient. Just remember to remove the pickup coil whenever you call the telephone company for service or repairs; the telephone company frowns on any special additions or modifications to their equipment, and the repairman will be completely within his rights to not only remove and confiscate your pickup coil, but possibly to remove your phone service as well.

PHONE RECORDING PROBLEMS

Generally, these are the problems in telephone recording: Poor signal level, which can be improved by using a more sensitive pickup coil or a better pickup coil location on the instrument; line hum, which might be caused in the tape recorder itself—especially if it is plugged into the ac line. If you're plagued by a hum problem, try reversing the plug in the ac receptacle, switching the machine to dc and recording with batteries, or moving the cable that leads to the pickup to the recorder. Generally, one or all of these methods will provide an improvement in the signal-to-hum ratio, and certainly will help to provide a better recording. Moving that cord around assumes that there is some stray ac pickup from the power line in the wall, or from fluorescent fixtures or other magnetic devices that radiate ac.

COIL IMPEDANCE MATCHING

A major problem at one time, impedance matching is largely taken care of with today's pickup coils. However, there are plenty of high-impedance types still around from the vacuum-tube tape recorder days, and you might just run into one. Impedance matching with a telephone pickup is just as important as it is with a microphone or a loudspeaker.

If you use a high-impedance pickup on the order of 10,000 to 50,000 ohms, you will get a very serious mismatch with the low-impedance input of your cassette tape recorder, resulting in a very low signal level. If you're not sure how old the induction pickup is that you're using, check the impedance by substituting another pickup coil or by measuring the impedance of the coil.

Sometimes an impedance mismatch may not be the problem at all, even though you're still getting a poor recording. It may simply be possible that your recorder does not have sufficient amplification at the input to do a decent job of recording the telephone conversation. If you suspect that this

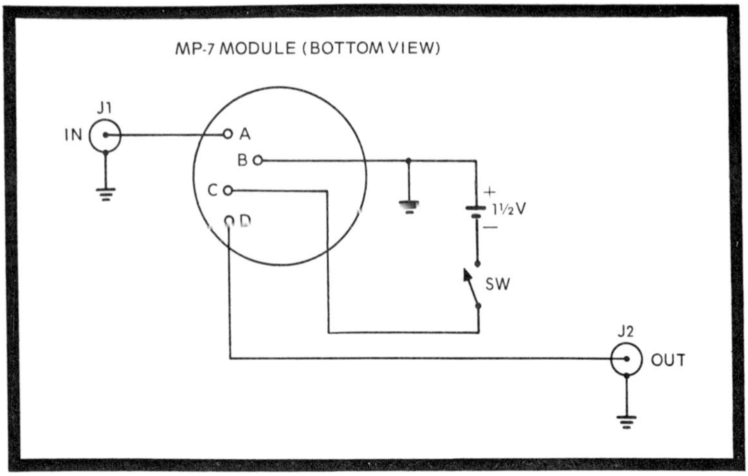

Fig. 14-4. Billed as a Citizens Band mic preamp circuit, this potted module from Carl Cordover can double nicely as a telephone preamp, giving lots of volume to drive even the least sensitive recorder. It also does a nice job of eliminating impedance-mismatch problems, too.

Fig. 14-5. A must in most telephone and other unusual recording situations is an adapter of some kind. This one accepts an RCA-type phono plug and terminates in a 3½ mm phone plug, which is accepted by most Japaneseportable cassette recorders.

is the case, then a preamplifier of the type shown in Fig. 14-4 can be made very inexpensively from a prewired, encapsulated circuit module. The use of a preamp is a good idea in all telephone recordings. It gives you the extra boost that you need for getting those "bad-connection" calls and certainly will provide enough amplification to overcome any impedance problems with the equipment that you're using.

One way to spot a low-impedance vs a high-impedance pickup is to check the kind of plug that it is supplied with. If it has the large-size standard phone plug (PL-55), it's a good bet that it is a high-impedance type, although this rule doesn't necessarily apply all the time. If it uses a conventional Japanese-type miniature plug with a 3½ mm diameter, chances are good that it's a low-impedance device and can be used directly with your tape recorder—provided, of course, that your tape recorder has this size jack. If the plug will not fit your tape recorder, you can use an adapter of the type shown in Fig. 14-5. One other point: Be sure to plug the unit into the **microphone** receptacle on your tape recorder. Otherwise, you're certain not to get a good recording, and possibly none at all.

AUTOMATIC EQUIPMENT

Possibly one of the greatest boons to business has been the automatic-answering device that answers your phone in your voice when you're out and takes messages. One unit of this kind is shown in Fig. 14-6. The cassette is a relatively recent entry into this area, since the continuous-loop cartridge has been the medium of choice until fairly recently.

The unit answers the phone with a recorded message that you prepare in advance. Some of the more sophisticated units will let you call your phone from some other location and dial a code number or tone to trigger the playback device. It will then play back all of your messages. Obviously, this type of convenience costs more money, and very few units are equipped with this special feature.

The unit can be connected to the line in one of several ways, although two are most commonly used. One way is to use a direct connection into the phone line itself, which allows the unit to answer the telephone electrically. An alternate method is by physical attachment to the telephone itself, by placing the phone handset on a special receptacle that fits over

A

B

Fig. 14-6. Automatic phone-answering system by Dictaphone uses the latest in cassette tape technology.

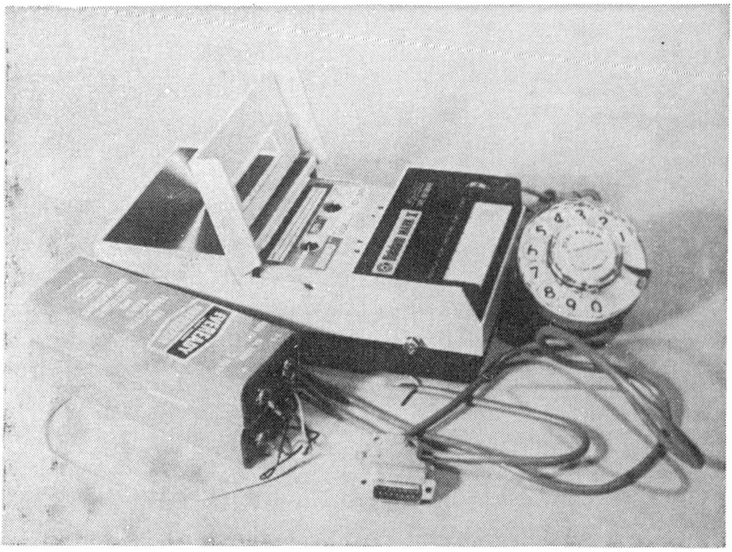

Fig. 14-7. The Dialalarm Mark I automatically dials the local police or fire department and plays a prerecorded emergency message. The telephone dial is an accessory used to encode the proper telephone number on the tape for automatic dialing. The unit is triggered by fire- or burglar-alarm sensors.

the telephone cradle. When the phone rings, the unit's cradle lifts the handset, and an induction pickup or a direct acoustical coupling to the handset connects the recorder, which answers the phone and takes the message.

Another system that is becoming popular is the automatically recorded message to call police or the fire department or both. With a unit such as the type shown in Fig. 14-7, the local police or fire department is dialed automatically when the unit is triggered by a burglar- or fire-alarm system. When the police or other public safety department answers the telephone, they will hear a recorded message something like, "This is a recording; there is a burglary in progress at 2703 Cedar Lane; please send a car at once." The message repeats and may be on an endless loop cassette of the kind shown in Fig. 14-8. The endless loop is probably the best method for such emergency messages, since it repeats the message several times or continuously until the appropriate public safety officers arrive.

One of the most important aspects of setting up an emergency cassette alarm is proper programing, and this is best done by following the directions that come with the unit. This is especially important, since each cassette tape system uses slightly different programing techniques, and there is no one general way to handle all of them.

Setting up the automatic dialer and alarm recorder requires proper attachment to some sort of triggering system; Fig. 14-9 is a block diagram showing how such an arrangement might operate. There are several such systems on the market, both ready-made and in kit form. There are also plans in various books and magazines showing how to build a system to suit your own needs. Whichever setup you elect to use, be sure the tape system makes a direct electrical connection with your telephone. The best way to do this is to have your local telephone company install a phone jack which makes the connection easy and foolproof, since all you have to do is wire your tape equipment into a plug and then run the telephone plug into another jack on the recorder.

Fig. 14-8. Useful for automatic telephone work, this TDK endless-loop cassette will work in most cassette recorders, and is available in lengths of one minute up to six minutes. At the end of the time, the message repeats.

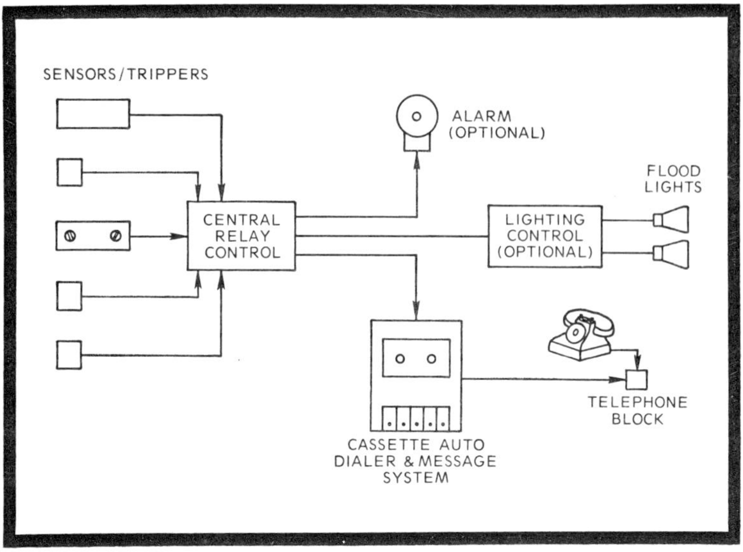

Fig. 14-9. Block diagram showing how a typical burglar alarm system is wired up with a cassette recorder and telephone. The recorder automatically dials the police or fire department and plays a prerecorded message announcing the nature of the emergency and your address. It does all this silently, so a burglar will continue his work while the police are on their way.

HOW CASSETTES ARE MADE

Chapter 15

The cassette is not just a length of tape; it is a very precisely cut ribbon of polyester, coated with ferric oxide, specially mounted on two hubs and installed in a relatively complicated plastic shell. Each end of the tape is spliced to a length of clear leader tape which is fastened to the two hubs. In between these two hubs, the tape passes over various rollers, guide pins, and a felt pressure pad. The cassette is actually part of the tape recorder; and, unlike open-reel tapes, it contains part of the tape recorder's mechanism.

The insides of a typical cassette are shown in Fig. 15-1. This illustration shows clearly the many parts that go into that plastic box. The more complicated the interior of the cassette, the better it usually is as far as construction quality goes. The parts shown in Fig. 15-1 represent probably the most quantity and engineering quality for such a design. Putting these pieces together is another matter.

ASSEMBLING THE CASSETTE

Preceding the basic assembly operation, the parts and pieces are molded, pressed, stamped, machined, and otherwise prepared. It is the hand assembly of these pieces into the cassette that is time-consuming and expensive.

The tape itself is slit precisely from large rolls measuring a yard wide (see Fig. 15-2), and from there the tape travels to some final assembly point where it ultimately is wound into the assembled cassette. While this is going on, the cassette is usually hand assembled—an operation that is more economical outside the United States because of cheap labor.

At least one cassette manufacturer has found a way of beating the high cost of American labor by developing a hulking monster of a machine (see Fig. 15-3) which assembles the cassettes automatically. This machine produces the C-0

Fig. 15-1. Components of a precision cassette: tape and leader on two hubs, Teflon slip sheet liners, flanged guide rollers, axle pins, felt pressure pad on a brass spring, mu metal shield, shell halves, and assembly screws. (Courtesy TDK.)

Fig. 15-2. This magnetic tape slitter cuts rolls of magnetic tape to correct width and lengths. (Courtesy 3M.)

Fig. 15-3. The first of a breed, 3M's automatic cassette assembly machine keeps nearly a dozen attendants busy loading parts and unloading finished cassettes.

(C-Zero) size cassette—the same type of cassette that hand-assembly operation produced previously. The C-0 is a cassette that contains a single piece of clear polyester leader tape clamped firmly to each of the two hubs and threaded through the guide posts and pins and in front of the felt pressure pad. Once this leader is installed, the other half of the plastic cassette sandwich is put into place and the package is ultrasonically welded together.

The tape itself is inserted into the cassette later. At that point, a tool is inserted into the cassette to pull out a loop of the leader tape. The leader is cut approximately in the middle of its length, and the end of the blank tape is spliced to it. A measured amount of tape is wound into the cassette, the tape is cut, and the newly cut end spliced to the other loose piece of leader. This process is shown in Fig. 15-4.

This technique is generally used by most blank cassette manufacturers. At 3M the process is taken a step further and is done automatically as the second step after the hulking monster finishes loading the cassettes with the empty hubs and making them into C-zeros.

With the rising cost of labor in Japan and other countries where cassettes are assembled, it seems fairly probable that in the near future; either that or find other countries with labor cheaper than their own.

PRERECORDED CASSETTES

The process of manufacturing a prerecorded cassette requires many of the same operations as the blank cassette, with one big exception—the tape that is used has been through a recording process before being loaded into the cassettes. Here's what happens:

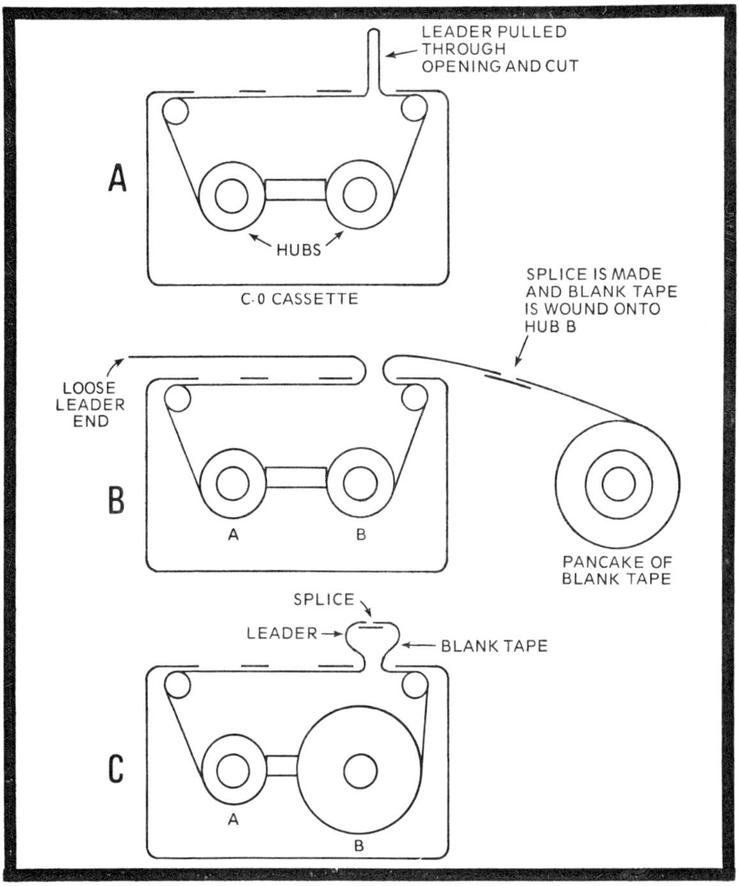

Fig. 15-4. During cassette assembly, a C-0 length of leader material is anchored between the two hubs (A). The leader is then pulled out and cut, the tape end is spliced to one leader (B). A measured amount of tape is metered into the cassette, then cut and spliced to other leader end (C). It's all done on an automatic machine.

Fig. 15-5. Prerecorded cassettes are mastered on these slaves from an endless loop of inch-wide master tape stored in the vacuum bin of the machine in the foreground. (Courtesy Ampex.)

A master tape of the material to be recorded—pop album, symphony, Broadway show, or whatever else it happens to be—is played at high speed on a master playback machine. This machine often is an endless-loop type as shown in Fig. 15-5. While this tape is playing at high speed, the narrow blank cassette tape unwinds from large blank "pancakes" on ten slave recorders, also at high speed. These slaves record the same material that is on the master tape. When the master finishes, it sends out a special signal tone which is recorded on each of the cassette tapes.

None of the equipment stops at this point, since the master tape is on an endless loop going through a bin which keeps it loose on a column of pressurized air and in a vacuum column. The machine simply plays the whole thing over again, and it's recorded again and again until the pancake is completely recorded. Each pancake is then lifted from its slave machine and transferred to the assembly line.

the end of each segment of tape. Each cassette is loaded with one segment of this recorded tape in special assembly stations of the kind shown in Fig. 15-6. At this point the tape is spliced into the cassette's leader and the metered amount is cut off—in this case, where the next signal tone is located—and the new cut end of the tape is spliced into the other piece of leader.

The loaded cassettes are then sealed together ultrasonically and are labeled and packaged for shipping.

There, the tape feeds through a machine not unlike a tape playback unit which senses the tone that has been recorded at even these countries will start to automate cassette assembly

Fig. 15-6. Huge "pancakes" of recorded cassette tape material are cut into individual cassette lengths and spliced onto hubs. (Courtesy Ampex.)

Fig. 15-7. For short-run duplication, preloaded blank cassettes are recorded individually from a master. This stack-loading Ampex duplicator can slave other machines from the master cassette.

Periodically, cassettes are picked out of this line for testing, and this spot-check quality control is important to the final product.

Another method of duplicating cassettes, which is used for short runs—typically a thousand pieces or less—employs one-to-one duplicators which run either at normal speed or possibly two or three times normal speed. One such duplicator is shown in Fig. 15-7. This one uses a premium-grade cassette for the master and accepts a stack of blank cassettes which it records one at a time.

Other types of machines, such as those made by Pentagon, will record up to four cassettes at a time from either a master cassette or a master open-reel tape. Unlike the Ampex machine, these units must be manually loaded and unloaded when the cassettes are finished. But a fast operator can service several of these machines, and can produce a cassette four times as fast as the single-cassette-at-a-time machines.

Short-run duplication of this kind is used to produce special courses of study that are recorded on cassettes, taped newsletters and magazines that are becoming increasingly popular, and other special purpose tapes that do not have a large, widespread market.

If you want to duplicate cassettes at home, it's a simple matter to set up two or three cassette machines side by side and run through duplicates on a one-to-one basis at the same speed. Unlike professional duplicating equipment, however, you must record both sides of the cassette separately by turning them over. Virtually all professional duplicating equipment records all four tracks at the same time by recording side A tracks in reverse as shown in Fig. 15-8. Almost any appropriate setup for home custom duplicating will do a superior job, but a lot of patience and time are required, and this is why using perhaps two or three machines connected as slaves would be preferable since this would save time. The problem with hooking up such slaves for home use is that paralleling the inputs to these machines will change their impedance characteristics, and this could result in a severe loss of signal level to the slaves.

DOLBYIZED CASSETTES

Some tape-duplicating companies, such as Ampex, are Dolbyizing all prerecorded cassettes that come off their production lines. Such cassettes, recorded on premium-grade tape, produce spectacular results when played back with

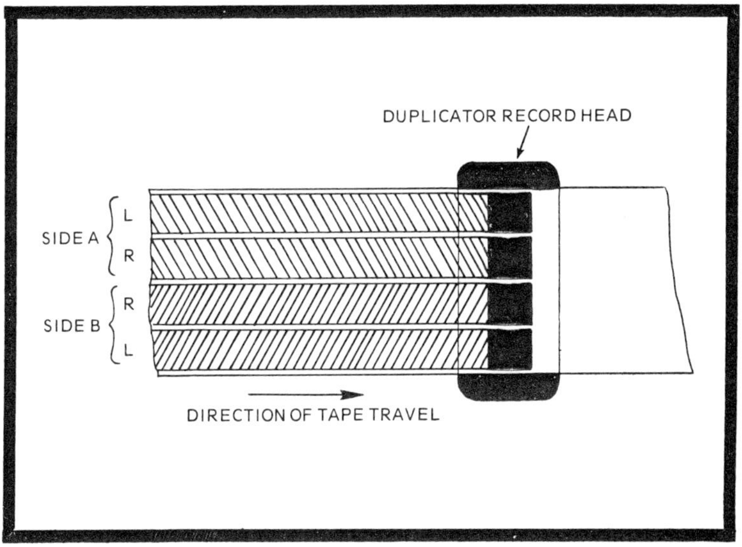

Fig. 15-8. All four tracks of a cassette tape are dubbed at the same time, with the two reverse-side tracks being recorded in reverse. This way, just one pass through the machine does it all.

Dolby-equipped cassette players. For listening on non-Dolby equipment, a satisfactory playback is possible by cutting down the treble control a little bit. If the treble control is left untouched when playing Dolby tapes under such circumstances, the music will sound a little too bright and harsh.

FOUR-CHANNEL CASSETTES

Chapter 16

The newest revolution in sterophonic sound is 4-channel stereo. It has a lot of names these days: quadraphonic, tetrasonic, quadrasonic, surround sound, and quad stereo. This development in the stereo medium is considered by anyone who has ears to be as much of an improvement over conventional stereo as stereo is over mono.

A convenient name for the new sound is "quad stereo," and it is being adopted by many segments of the industry. What it does is to help recapture the lost sound field—the total acoustic effect, reflected sound and reverberation present in any live performance. If you have ever tried to reproduce the sonics of a Carnegie Hall in your living room, you may have noticed that your listening room is simply not up to the task. This is because your room is vastly different in size and in acoustic qualities. One way of getting around this problem is to reproduce the effect of the large-hall acoustics by recording these reverberations, phase changes, and other phenomena on two additional channels which are played back through two speakers placed at the rear of the listening room. These two rear-channel speakers can be located in one of the arrangements shown in Fig. 16-1.

The 4-channel effect, of course, does not have to be limited to material recorded in a concert hall. With a total 360-degree circle of sound—known as the "sound field"—it's possible to create many unusual and exciting sonic effects with almost any kind of music. There are even works in classical literature that call for musical forces located in the balcony or rear of the listening hall. Just try to reproduce the directionality of something like that with ordinary two-channel stereo! With quad stereo it all becomes possible. It's quite an experience to hear a brace of trumpets blaring out from the rear of the room when least expected.

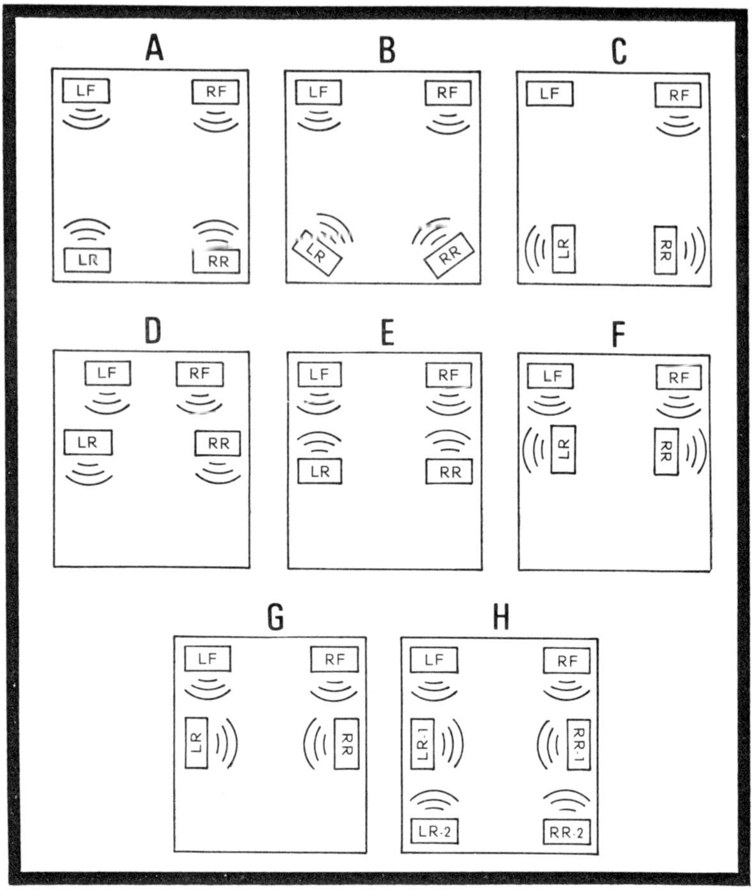

Fig. 16-1. Some of the many possible speaker arrangements for listening to 4-channel stereo. The actual arrangement you use depends largely on your own musical preferences. Notice that H is a combination of A and G and, although it uses a total of six speakers, it can give the most satisfying results.

RECORDING QUAD

The basic process for recording and reproducing quad stereo is shown in Fig. 16-2. For this classic method, four tracks that cover the entire width of a ¼-inch tape are recorded and reproduced individually through four separate amplifiers and speakers. This method of reproducing quad stereo is called "discrete" and is referred to quite a bit because it is supposed to be the ideal form of quadraphonic reproduction.

While the discrete method certainly provides the best possible results, it is by no means the most practical, since it cannot be broadcast without major changes in station equipment and in FCC regulations. The equipment involved may be very costly and there is a very limited selection of prerecorded discrete tapes.

In the quest for a more convenient and lower-cost method of reproducing quad stereo, both cassettes and cartridges have entered the picture. The very first popularly priced consumer product for 4-channel reproduction to hit the market uses an ordinary 8-track cartridge, but it offers only half of the normal playing time, since there is twice as much material for each program. This format was introduced by RCA in 1970, and they immediately started to produce a substantial number of what have become known as "Quad-8" cartridges. There are several manufacturers now who make equipment both for cars and home use that accept these cartridges. The introduction of quad has certainly prevented the cartridge from dying an early death in favor of the cassette.

FOUR-CHANNEL CASSETTES

The cassette is quite another matter. There are two major theories involved with discrete cassettes. The obvious method

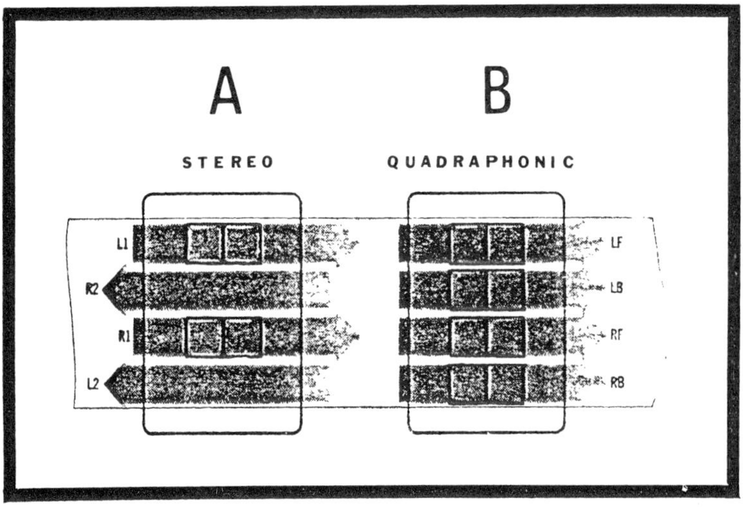

Fig. 16-2. Open-reel tape recording format for 4-track, two-channel stereo (A) and 4-channel stereo (quadraphonic).

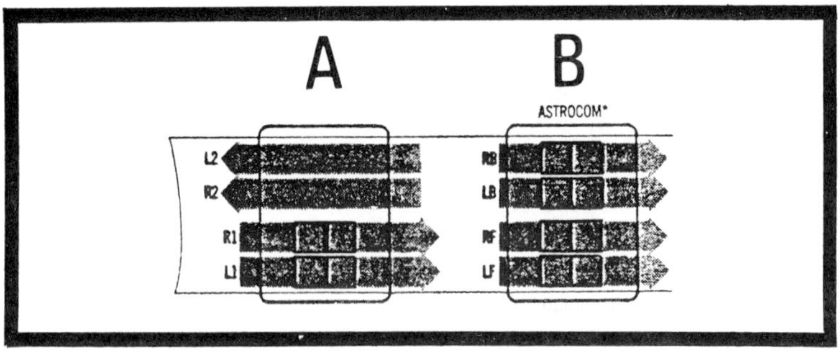

Fig. 16-3. Recording format for cassette tapes: (A) 4-track, two-channel stereo; 4-channel discrete stereo, using the same track width as two-channel recordings. This quad arrangement is advocated by Astrocom/Marlux and other manufacturers.

is to use all four tracks of the conventional stereo configuration for quad as shown in Fig. 16-3. This method has two obvious advantages: it does not obsolete present equipment that can play back just the two front channels, and it can be produced with existing heads and recording equipment. This method does, however, eliminate cassette turnover for second-side playback; the cassette can be played in one direction only.

But Philips wants it another way. Since any standards that Philips sets must be followed by the entire cassette industry, there seems to be somewhat of an impasse right now. The track arrangement that Philips would like to see used is shown in Fig. 16-4. This track arrangement follows through with the total Philips concept of making everything on cassettes compatible with everything else. This means that any 4-channel material recorded on a cassette can be played back on two-channel stereo equipment and even on monophonic equipment, while at the same time the cassette still provides second-side playback. So far, the only actual cassette machine in production for 4-channel use, uses the track configuration shown in Fig. 16-3. This machine, made by Astrocom, is shown in Fig. 16-5.

The major problem with the Philips method is a technical one. The signal-to-noise ratio of the already narrow track width is bad enough. Cut this width by more than an additional 50 percent, and the signal-to-noise ratio of any given channel must suffer enormously. Advocates of the Philips system can point out that the overall signal-to-noise ratio is not that noticeably different, since you still have an approximately

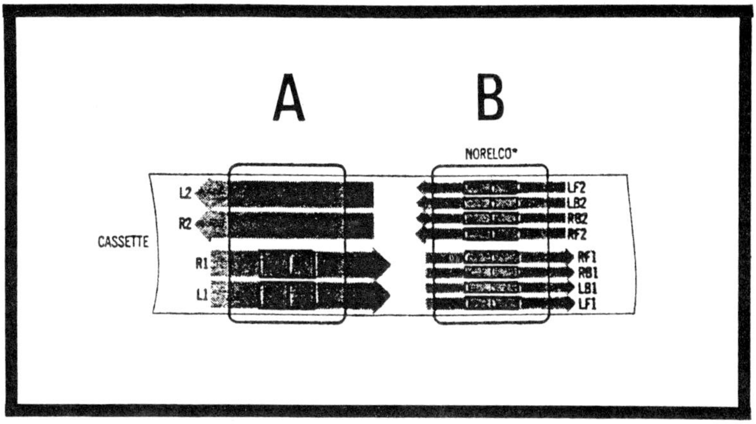

Fig. 16-4. Proposed Philips format for cassette tapes: 4-channel, 8-track discrete quad. This track arrangement would provide total system compatibility, but would greatly degrade the signal-to-noise ratio. The format on the left is used for two-channel stereo.

Fig. 16-5. Astrocom/Marlux 307 discrete quadraphonic cassette deck, using the track arrangement shown in Fig. 16-3 B. Unit both records and plays back quad.

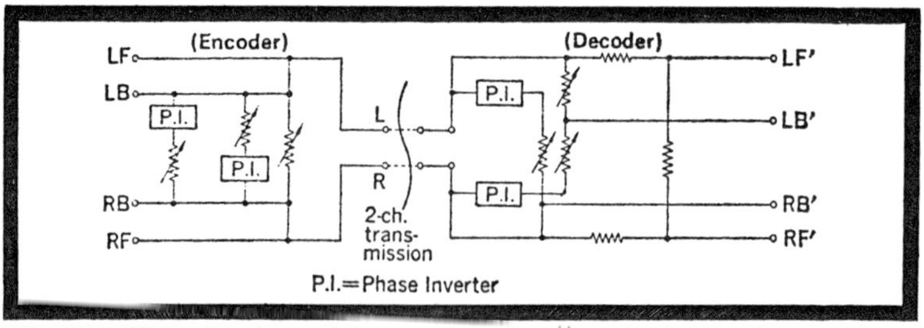

Fig. 16-6. This simple matrix encoder circuit mathematically combines rear-channel signals into the front channels for two-channel recording.

equal overall signal level for the entire system. But noise measurements, unfortunately, are taken one channel at a time, and this is where the system suffers. Another problem with the Philips system is the design of heads that are up to the task of recording and reproducing such an extremely narrow track. Such heads do exist and can be improved upon, but they are expensive and difficult to make at this point.

MATRIXED QUAD

The ultimate answer probably lies in the method that has been developed for recording quad stereo on records. The mind may boggle at the thought of putting four channels of sound on just one record groove, but remember that it was thought impossible to put two channels of sound in one record groove not too many years ago. But matrix techniques have already been very successful. Not unlike the matrix used for broadcasting stereo on a single FM station, a typical encoding system in its simplest form for providing this matrix is shown in Fig. 16-6. The mathematics of this system are as follows:

$$L = (LF + LB)\cos\theta + (RF - RB)\sin\theta$$
$$R = (RF + RB)\cos\theta + (LF - LB)\sin\theta$$

Then, the relative decoder output will be:

$$LF' = L\cos\theta + R\sin\theta = LF + 2RF\sin\theta\cos\theta + LB\cos2\theta$$
$$RF' = R\cos\theta + L\sin\theta = RF + 2LF\sin\theta\cos\theta + RB\cos2\theta$$
$$RB' = R\cos\theta - L\sin\theta = RB - 2LB\sin\theta\cos\theta + RF\cos2\theta$$
$$LB' = L\cos\theta - R\sin\theta = LB - 2RB\sin\theta\cos\theta + LF\cos2\theta$$

From the above equations, if LB=RB=1, or when there is a sound source at the back center of the original sound field, the following equations are obtained:

$$L = LF\cos\theta + RF\sin\theta + (\cos\theta - \sin\theta)$$
$$R = RF\cos\theta + LF\sin\theta + (\cos\theta - \sin\theta)$$

This whole idea may sound a little bit weird, but it really works. What it does is condense four channels of sound into two channels. The four channels are then unscrambled by a decoder at the playback end where they become four full channels again. Naturally, some of the front-to-back channel separation is lost in the process, but not enough to make any significant amount of difference. The system works so well that it can be broadcast on any FM stereo station without any special equipment or trouble with the FCC.

Quad can also be encoded into two normal channels of sound on any tape that is manufactured today. This system is totally compatible, and a full two-channel stereo program is obtained on playback with ordinary stereo equipment, just as a full monophonic program is obtainable on monophonic cassette players.

DECODING MATRIX

A typical decoding system for matrixed material is shown in Fig. 16-6. Such a decoder can be made very simply or somewhat more costly, depending on the basic matrix designed into the system. Some companies are offering a simple matrix decoder for as little as $20 in kit form (see Fig. 16-7), while other units can range up to $100-plus for decoders in combination with rear-channel amplifiers.

Most encoding systems cause a certain amount of phase cancellation in the two rear channels because they must be 180 degrees out of phase with each other in order for them to be matrixed properly on the disc. To avoid this cancellation, the Sansui QS system processes the two rear channels before encoding and rotates each one of them 90 degrees from the original position, providing a plus 90-degree shift for the right rear channel and a minus 90-degree shift for the left rear channel (see Fig. 16-8). This places the two rear channels 180 degrees out of phase with each other, but without causing any cancellation in the actual disc-cutting process.

The major matrixing systems are largely compatible; a quad recording made with one such system can be played back using another system's decoder with good results. At this writing, there are two major encoding systems in use—the CBS SQ system and the Sansui QS, both of which are compatible with each other.

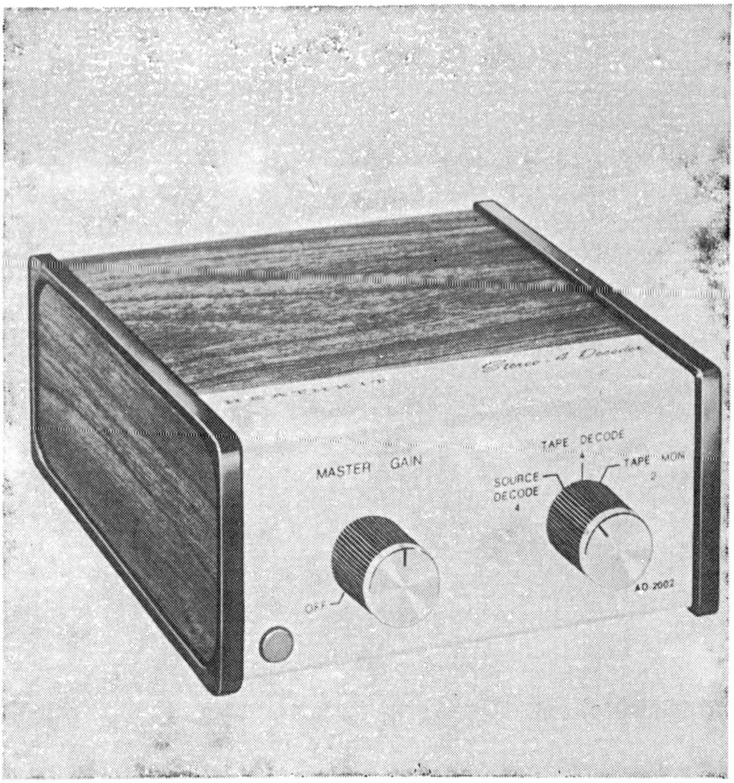

Fig. 16-7. Inexpensive 4-channel decoder made by Heath does not accommodate differences in competing matrixes, but will work well with all of them.

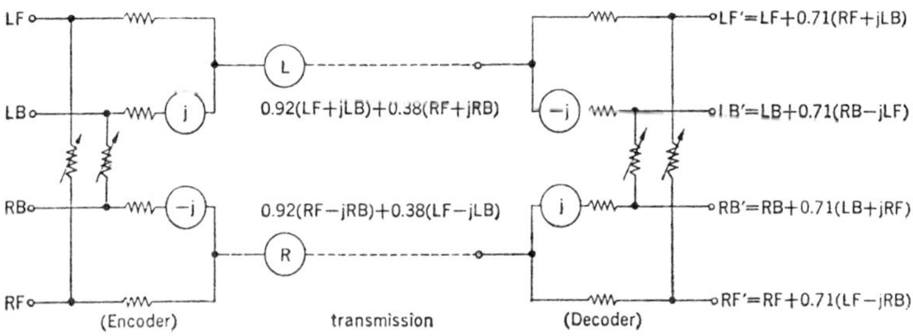

$$\theta=\frac{\pi}{8}\quad\begin{cases}\tan\theta=0.414\\ \cos\theta=0.92\\ \sin\theta=0.38\\ 2\sin\theta\cos\theta=\cos 2\theta=0.71\end{cases}$$

Fig. 16-8. Block diagram of the Sansui QS regular matrix encoder-decoder, now the standard in Japan. It may become the worldwide standard eventually.

Another system which has begun to wither on the vine is the RCA/JVC technique. This is not a matrix system at all, but a multiplexing system which requires extremely critical recording and playback equipment. Very special care must be taken in handling and playing back the discs.

Ultimately, the matrix will be in industry-wide use. It does not require any special equipment at the broadcast station and no special equipment need be used with conventional tape recording machinery. If you want to enjoy full quad playback from a matrixed tape, all you have to do is insert an inexpensive decoder between the tape deck and the amplifier and add another two-channel stereo amplifier and speakers.

If this all sounds too simple, just remember how basically simple matrix quad is. It's just an extension of single-groove stereo recording, and after all, time was when that, too, was considered technically impossible. And discrete 8-track cassettes? Not feasible at this moment, but then neither was the compatible stereo cassette a few short years ago. Time will tell.

SERVICING PORTABLE CASSETTES

Chapter 17

The battery-operated portable recorder is the most prolific type of recording instrument in the world today. There are millions of them in the field, and they are sure to need servicing from time to time. The typical battery portable—in fact the archtype and prototype for all of them—is the classic Norelco 150, which was the very first cassette recorder. This model has been superseded by the Model 1420, which is just as typical and an all-encompassing type of recorder.

Probably one of the most difficult initial steps in servicing any portable recorder is getting it out of the case. First step in all portables is to remove the batteries and any loose parts, covers, etc., associated with the batteries. After that, go through the machine and remove all surface Phillips-head screws; be sure to use a jeweler's-size Phillips screwdriver, since most of these screws have very tiny slots that will be damaged by using one of the more conventional size tools (see Fig. 17-1). Be sure also to remove any knobs that might interfere with the case removal. Most portable control knobs are friction fits and can be pried loose; just take care not to damage the case or the knob while you're prying.

CLEANING & LUBRICATING

One of the primary causes of equipment failure is a lack of proper cleaning and lubrication. While this may seem self-evident, most cassette recorders do not get the kind of periodic maintenance that they require. Certainly, the owner may keep the heads and capstan and pinch roller clean, but the interior mechanism needs much closer attention. Most manufacturers recommend cleaning and lubrication every 500 hours of playing time. If the average cassette recorder gets this kind of attention that often, it's indeed unusual.

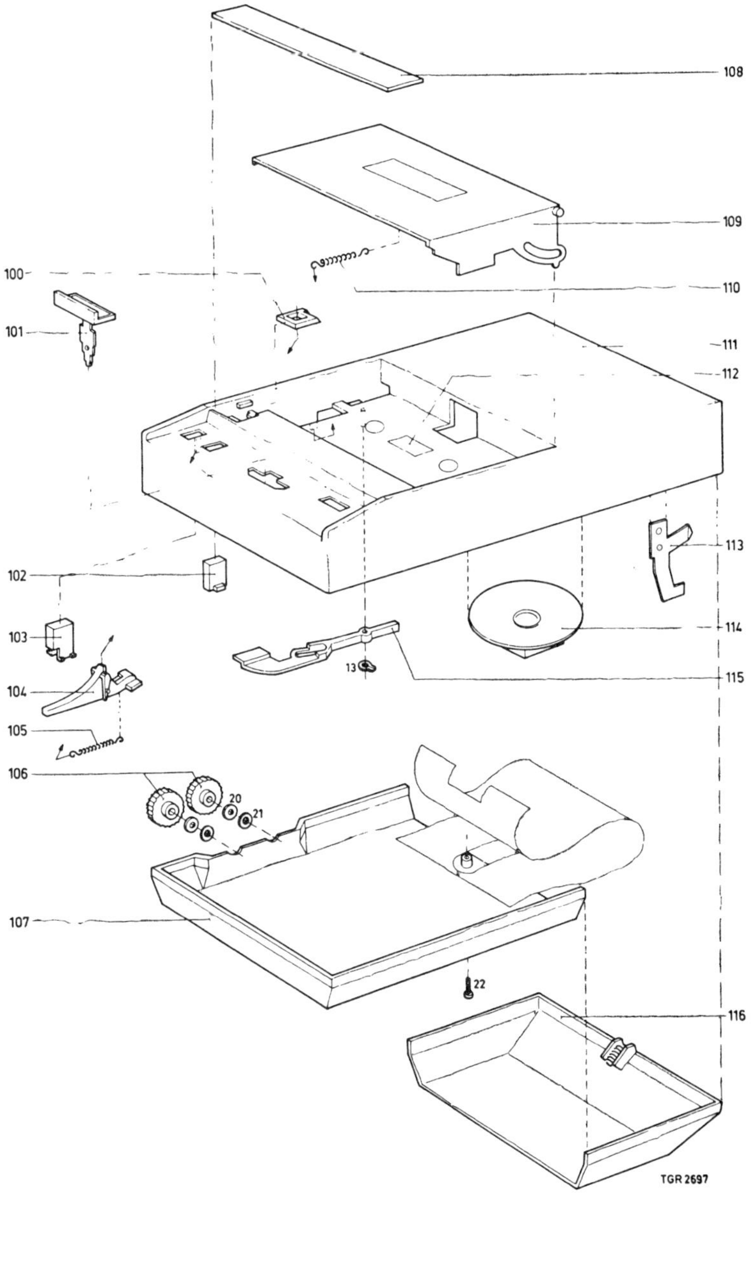

Fig. 17-1. This exploded view of a typical portable recorder shows case disassembly. (Courtesy Philips.)

While many prepared head cleaners are available, some manufacturers strongly urge that you use only alcohol for head cleaning, since some commercial cleaners might react unfavorably with encapsulating compounds and other sealing chemicals used in the head assembly.

Lubricant should be applied to the grooves and extrusions in all sliding subchassis elements. A good lubricant for this type of operation is a paste material such as Lubriplate. A high grade machine oil should be applied to the friction points for the turntables, the roller spindle, the flywheel spindle, the hub and bearing of the friction wheel, and the hub and spindle of the pulley. In all such cases be extremely careful, when lubricating, not to get any oil or other lubricant on friction surfaces such as pulleys and belts. A glance at the exploded view drawing in Fig. 17-2 will give you a rough idea of the many points that should be lubricated during this maintenance procedure.

Exploded views of this kind are usually part of service manuals available from various recorder manufacturers. It's a good idea to keep a file of service manuals on equipment that you feel will be crossing your bench regularly. If you handle this equipment as a dealer, then you definitely must keep such manuals on hand.

DRIVE MECHANISMS

Once you open up a recorder, plan to replace the rubber drive belt, since this is a prime source of trouble. As with all rubber parts, this belt tends to crystallize and harden with age, and chances are if the recorder isn't showing signs of such trouble now, it will be very soon—especially if it's in your shop.

Drive belts in cassette recorders look like very little more than a round rubber band, and some manufacturers use exactly that for drive belts. Exact replacements are available either from the manufacturers or from catalogs of general replacements, such as those produced by the Walsco division of GC Electronics.

To replace the belt, loosen the screws that hold the lower bearing bracket on the flywheel to the chassis. Generally, this will allow you to remove the washers and the bearing bracket itself. In some machines, you'll have to remove the mounting

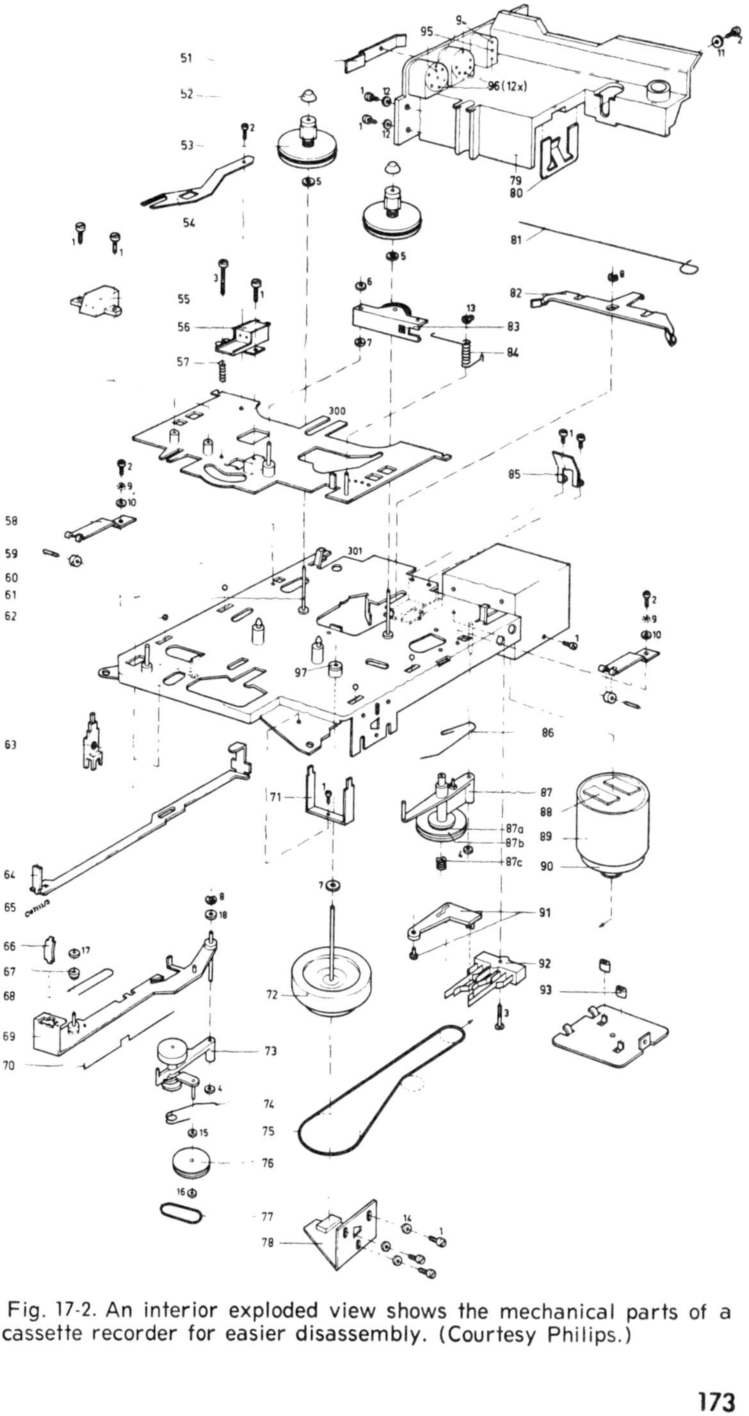

Fig. 17-2. An interior exploded view shows the mechanical parts of a cassette recorder for easier disassembly. (Courtesy Philips.)

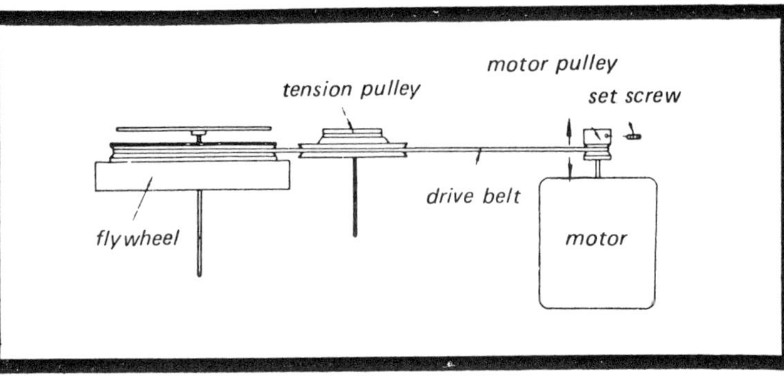

Fig. 17-3. Typical drive belt assembly. (Courtesy Sony.)

screws entirely to do this. If there is a separate cover plate on the motor housing obstructing the belt, remove this also (see Fig. 17-3). In most cases this is all that is necessary to get the drive belt out of there. In some machines it will be possible to remove the drive belt simply by loosening the flywheel bearing bracket and backing it off far enough so that there is a space between the bracket well and the bearing itself.

When replacing the drive belt, be sure that it doesn't come into contact with the bearing when you slip it over the bearing, since any lubricant on the bearing will contaminate the belt and cause it to slip. If your attempts to prevent this fail, and you do get a little smear of lubricant on the belt, clean it off with alcohol and let it dry before continuing to assemble it. The belt itself slips over the flywheel and the motor pulley, and there should be a fair amount of tension on the belt when this is done. Replace all the mounting screws, tighten them into place and turn on the motor to make sure that everything is operating properly.

SPEED & HEAD ADJUSTMENTS

The actual tape speed is another factor and, with many portable units, cannot be easily adjusted; however, some units such as the Norelco Carry-Corder contain an electronic motor-speed control circuit (see Fig. 17-4), which can be changed with a simple screwdriver adjustment. The way to check on motor speed—certainly the easiest way—is to use a specially prepared timing tape made by Philips. This tape contains a 6300 Hz tone over its entire length and has a half-second-long marker of 800 Hz that occurs every 100 seconds.

To check the recorder's speed with this tape, simply measure the time between any two of those 800 Hz markers; it should measure somewhere between 97 and 103 seconds for a tape recorder that is properly adjusted. Anything beyond these limits requires some kind of change. Some manufacturers offer an optional method of adjusting motor speed—installing a different motor pulley, for example. SONY offers four different pulleys for a typical machine, only one of which is classified as "standard." The other pulleys provide tape-speed differences of −2 percent, −1 percent, and +2 percent. Obviously, this is a relatively inexact way of adjusting tape speed, but it can bring the speed within the required tolerances for normal use.

Adjusting the play-record head azimuth is a relatively simple matter on most tape recorders. There is a spring-loaded adjusting screw, usually on the feed hub side of the tape head (see Fig. 17-5). By turning the screw, it's possible to reset the tape azimuth within acceptable tolerances. Azimuth is noticeable as a problem when there is an excessive amount of crosstalk between monophonic channels. Actually, there should be no crosstalk at all, since we're talking about a half-

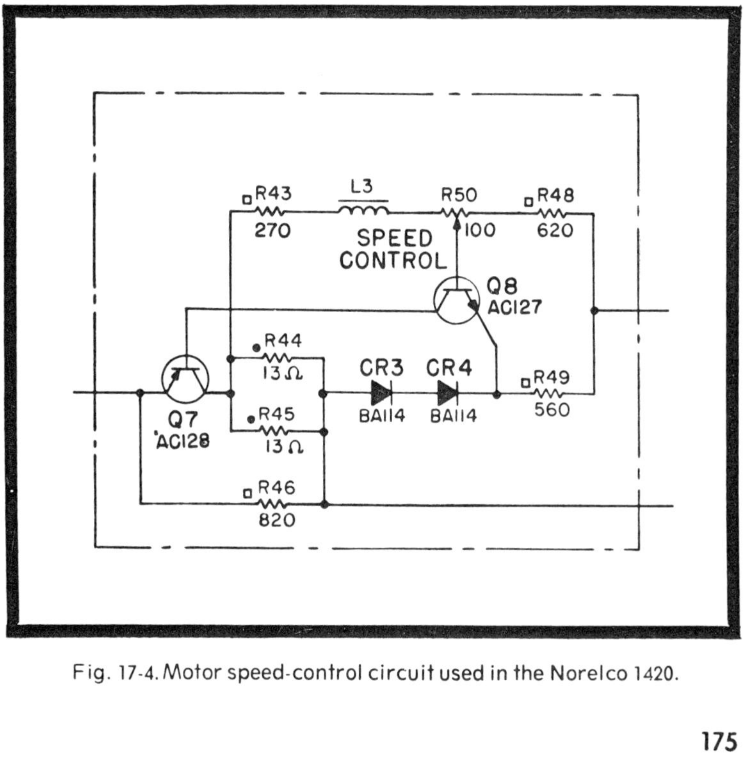

Fig. 17-4. Motor speed-control circuit used in the Norelco 1420.

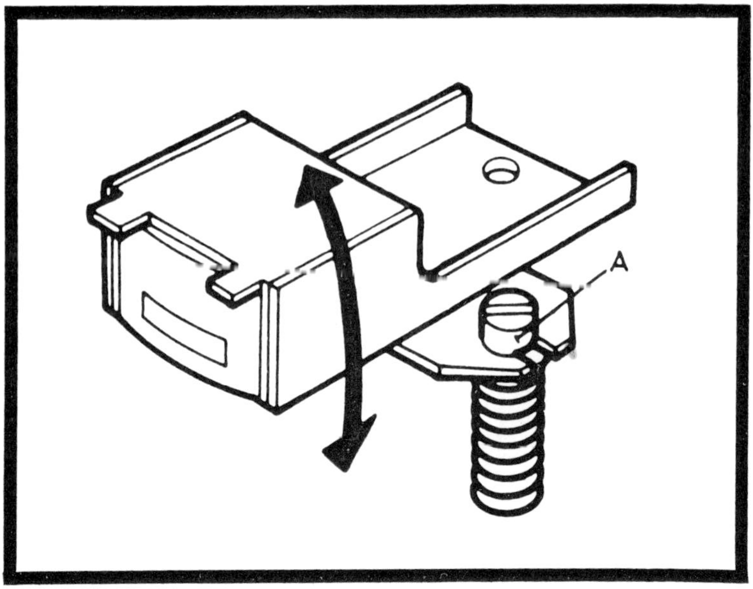

Fig. 17-5. Record/playback head azimuth adjusting screw in a Norelco.

track head which is well within the width prescribed by the tape. After adjusting this screw, it should be doped with nail polish or some other plastic compound that will dry hard to prevent its turning accidentally.

PRESSURE ROLLER ADJUSTMENT

One source of problems can be the amount of tension provided by the pressure roller or pinch roller against the capstan. This rubber roller must pinch the tape tightly to the needle-thin capstan in order for the tape to be pulled through at the proper speed. One way of measuring this is with a tension scale with fine calibrations. The force necessary to pull the pressure roller assembly clear of the capstan should be somewhere between 5.3 and 6.35 ounces. If the pinch roller is exerting either too little or too much force, it can be adjusted in Norelco machines by repositioning the tension spring in one of several holes provided for it (see Fig. 17-6).

In other machines, the tension can be measured the same way, but should be a different amount. A spring scale attached to the pinch roller in a SONY machine, for example, should be somewhere between 9.9 and 12.7 ounces. This measurement is made with the machine in the forward (play) mode with a scale attached as shown in Fig. 17-7. With the spring scale attached as shown, the scale should be pulled to separate the pinch roller from the capstan shaft. A reading should be taken when the pinch roller just begins to rotate again with a slight release of tension on the spring scale—just enough to barely make contact with the capstan shaft. The enormous difference in the respective pressure or tension ranges for the Philips and SONY machines indicates that there is considerable latitude in design philosophy among machines that are all intended for the same thing—battery portable operation. Again, the service manual appropriate for the recorder you are working on is very important to doing the proper job.

TURNTABLE REMOVAL

There will be times when you must remove the turntable from a portable machine to gain access to crucial parts. On all machines the cap at the top of the turntable shaft must be removed first. For some machines, such as the Philips series, the cap is a friction fit and can be pulled straight off; then the

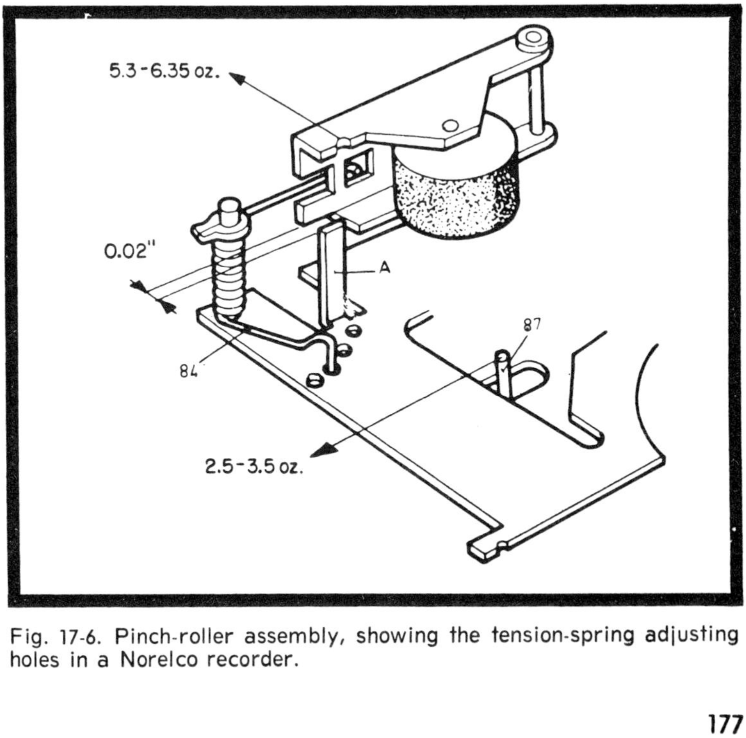

Fig. 17-6. Pinch-roller assembly, showing the tension-spring adjusting holes in a Norelco recorder.

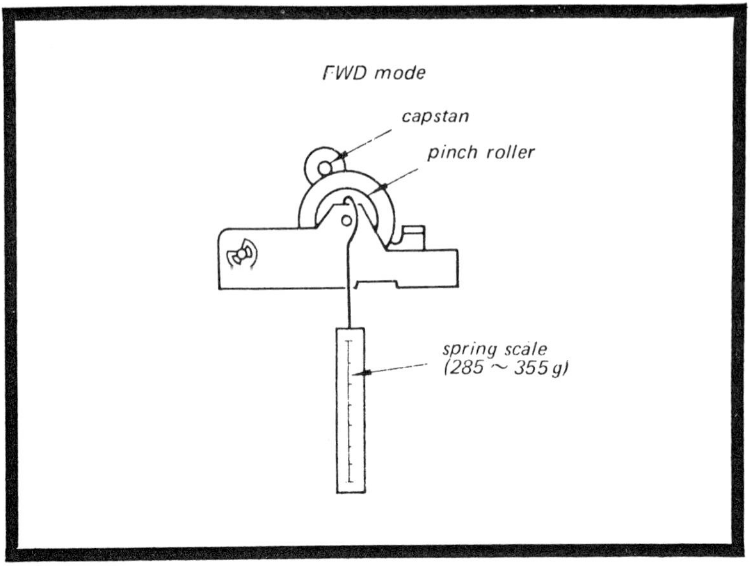

Fig. 17-7. Method of using a spring scale to measure pinch-roller tension. (Courtesy Sony.)

turntable can be removed. In all cases, when removing the turntable and reel spindle assembly, be sure to look for any fiber washers or metal washers that are below the turntable on the axle shaft. Very often these will stick to the turntable when it's removed and may fall off later before reassembly. Double-check on these washers, because they are almost impossible to replace without going back to the factory.

Sometimes the turntable needs replacement because the rubber tire along its edge has become worn or crystallized. This is an extreme case, but sometimes is the only type of surgery that will repair the machine. Be sure that the turntable rubber is definitely at fault before replacing it, since this is a lengthy operation—ordering the parts and holding the machine aside while you're waiting for them to come in.

ELECTRONICS SERVICING

The electronic portion of a portable recorder generally requires very little servicing unless the machine has been subjected to extreme abuse. The electronics is fairly straightforward as shown in Fig. 17-8. As you can see from this illustration, the single playback-record head is connected to a series of amplifiers and controls, ending up in a push-pull audio output stage which feeds the loudspeaker and extension or earphone jack. The only real confusion in reading a tape recorder schematic is in the switching and the switching functions, which have been simplified in Fig. 17-8.

In Fig. 17-9 the electronics for the record function is shown in simplified schematic form. The same transistors are involved, although here the signal source is a microphone instead of the playback head, and there is the additional feature of a bias and erase oscillator which feeds both the erase heads and record head (which is a playback head in Fig. 17-8).

Generally, electronic circuit problems occur during the record function far more often that they do in playback. One of the reasons for this is the possible misadjustment or lack of proper operation of the bias and record oscillator. Generally, it is a good first step with lack of recording or bad recording to check the ac bias voltages and possibly replace the appropriate capacitors—usually the bypass in the base circuit and in the transformer line. These two capacitors are almost always suspect where oscillator problems occur.

General symptoms such as a lack of playback could result from a defective or dirty record-playback head or possibly an open circuit to the head. The shielded lead going to the head is also a source of trouble and should be checked carefully with a ohmmeter. Naturally, the power sources should all be checked carefully and voltage checks made before digging in too

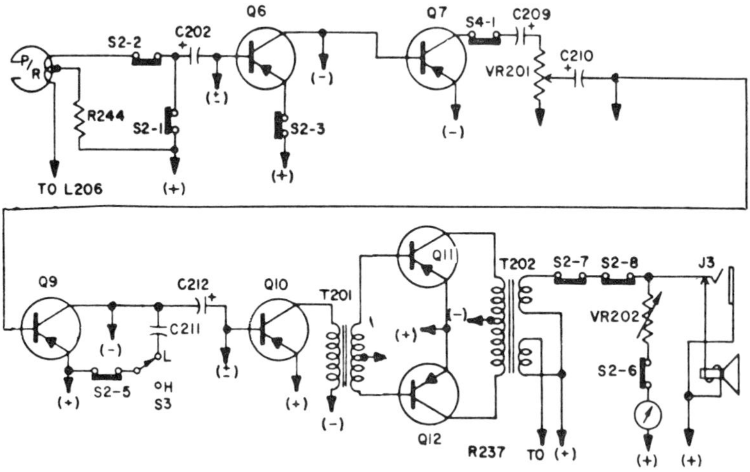

Fig. 17-8. Schematic of the playback signal path in a mono portable recorder. (Courtesy RCA.)

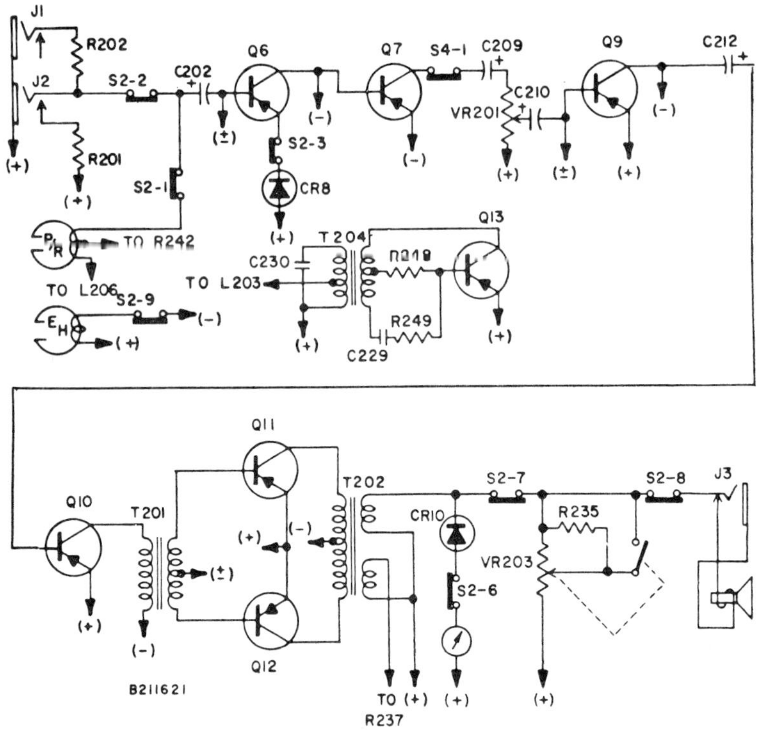

Fig. 17-9. Record signal path in a portable recorder. (Courtesy RCA.)

deeply. Sometimes the diode or the function switch can be defective. Before tearing out the switch, try a spritz of contact cleaner. It works for TV tuner controls, and it can be a big help here, too.

If there is a low playback level or distorted playback problem, again check the head, and also check the diode feeding the base of the second preamplifier. Another possibility here is that there might be some defect in the amplifier circuit itself, but this would also show up with a lack of recording as well.

A common complaint among cassette owners is that the record interlock button can't be depressed. If you find that this complaint isn't justified, be sure to ask to see the kind of blank cassettes your customers are using for recording. Chances are they may have broken out the interlock tab or are using a second-hand cassette with the interlock tab missing and don't know the difference. A lot of servicing problems can be solved just by looking at the cassette they're trying to use.

If the machine doesn't record, the record interlock lever itself could be defective. Again, check the record-playback head. If it is defective, of course it will also show up in playback. Where a machine will not record properly but still plays back, the prime suspect again is the oscillator and here you might also want to check the coil.

We have also seen microphones go bad somehow or other even though the dynamic microphones used with today's cassette recorders are supposed to be pretty nearly indestructible. The microphone cable and jack tend to wear, and this might be a major problem area. Be sure to check the cable and plug for intermittent operation.

If a machine records but doesn't erase, then you might nail the trouble down to a defective erase head or possibly one of those capacitors in the oscillator circuit. After all, if something doesn't erase properly, it probably isn't going to provide the proper recording bias either, since both come from the same oscillator.

Hum is usually a problem only when there is ac pickup nearby or if the machine is used with an ac adapter. This is a case where there is some leakage and possibly bad filtering in the adapter, in which case capacitors should be checked carefully.

Generally, good common sense and a good set of servicing instructions are really all you need to handle 99 percent of all portable cassette recorder servicing problems. These machines are relatively simple and should present no more of a challenge than their open-reel predecessors. Still, because of their compactness, there is a certain amount of reluctance to dig into them. They're actually easier to service than most transistor radios. After all, it's all laid out there for you like a mechanical schematic.

SERVICING CASSETTE DECKS

Chapter 18

Once the cassette was recognized as a potential high-fidelity medium and started seeing service for stereo use, an entirely new set of specifications and problems arose. As a home entertainment instrument, it's no longer just enough for a cassette machine to operate and to provide a somewhat reasonable playback. Now, it must also provide high fideity, and this brings with it all of the usual problems of high-fidelity reproduction with a few more thrown in—how do you squeeze good fidelity out of that tiny cassette even with the latest in technology?

HIGH-FIDELITY CHECKS

One way to achieve high fidelity is to add all kinds of fancy new electronics. With the introduction of chromium dioxide tapes, the cassette machine required an additional bias circuit, which is shown in Fig. 18-1. Also shown in this illustration are such special features as a Dolby noise reduction circuit, and other esoteric circuits.

When space-age circuits such as a Dolby are used, additional calibrating functions must be added to the recorder for setting these circuits at their proper levels. There is a recording calibrating tone, a chromium dioxide record calibrate circuit, and special calibration in the Dolby itself. Both the Dolby and the chromium bias circuits must have bypass and defeat switching, and the bias itself must be rock solid.

When we start talking about really expensive high-quality decks, there is some additional circuitry, usually in the form of a special servo drive for the motor. This is a special electronic circuit which keeps the motor running at a rock-steady speed, and in general reduces the wow and flutter component

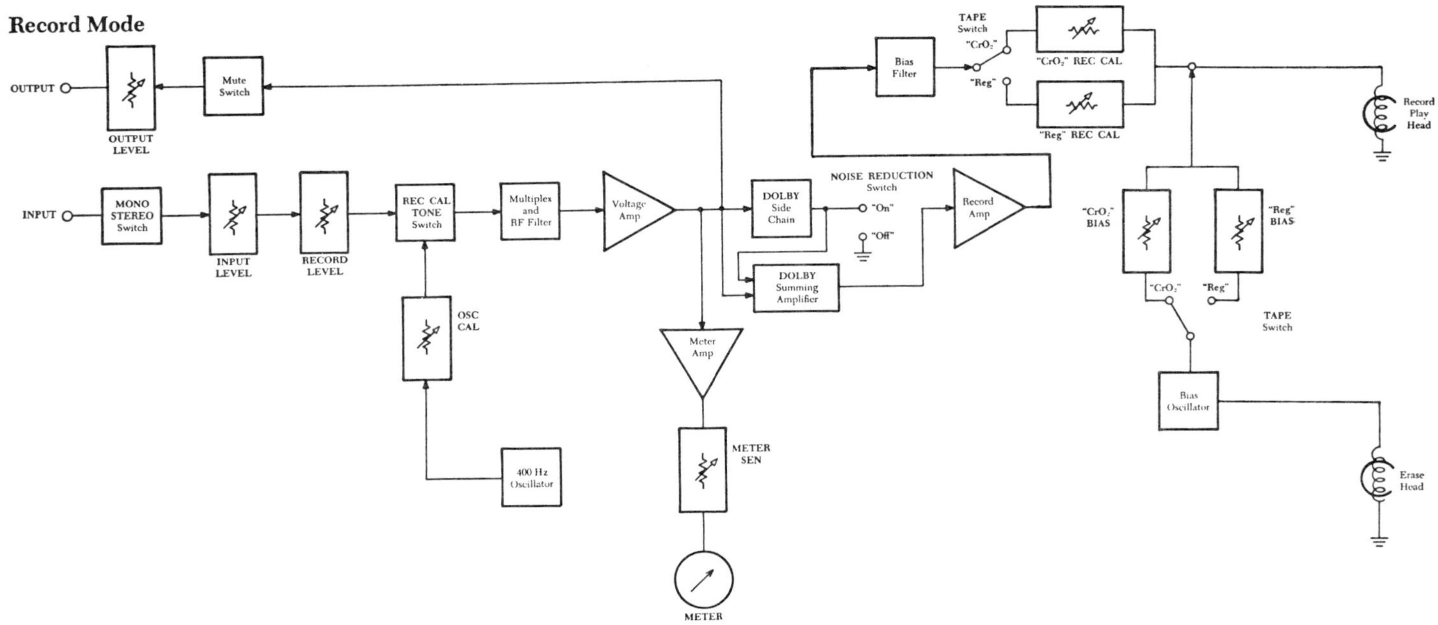

Fig. 18-1. Block diagram of one channel of a professional-type stereo cassette deck—the Advent 201.

of the entire cassette system. A servo drive of this type is shown in Fig. 18-2. Also in this illustration is an automatic shutoff system which both turns off the motor and releases any of the depressed function control buttons. If this machine is in the play or record mode, the buttons will pop up automatically at the end of the cassette. The same is true if the cassette is in the fast forward or rewind mode; the tape will stop and the buttons will pop up. The service manual provided for the recorder shows voltage test points for checking such circuits. Notice that in the motor servo section there is also an adjustment to touch up the motor speed control.

The electronics for a stereo deck is somewhat more complex than for a monophonic voice-only unit. Figure 18-3 shows a complete schematic diagram for a typical deck that does not have any of the niceties such as Dolby and switchable bias. Even though no power amplification is needed to drive speakers, each channel of record and playback requires five transistors plus a diode, along with a full-wave power supply and a two-transistor bias and erase oscillator circuit. Notice that TR11 and 12 (the oscillator circuit) operate through a transformer that couples both to the erase head and to the record circuits feeding the record head. This type of circuit is probably most common among stereo cassette decks; but because it is so straightforward, it is probably the least troublesome. There are no fancy circuits such as found in the audiophile decks that require touchup adjustment and other special treatment.

MECHANICAL SERVICING

Removing the case from a stereo deck is relatively easy, and about the same process is required as for most high-quality stereo components. The first step is to remove all knobs, which by and large are held in place by friction fit. A few may have set screws to be loosened. Once the knobs are removed, turn the unit over. Place it on a soft cloth to prevent scratching any surface or plastic parts, and remove or loosen all obvious case mounting screws. Usually, these will be Phillips-head screws, and there will be four or six of them on the bottom of the case and possibly a couple more on the sides. With these screws removed, the works should lift right out of the shell, exposing the upper portion. An additional set of screws might hold some retainer or mounting plates on the machine (see Fig. 18-4).

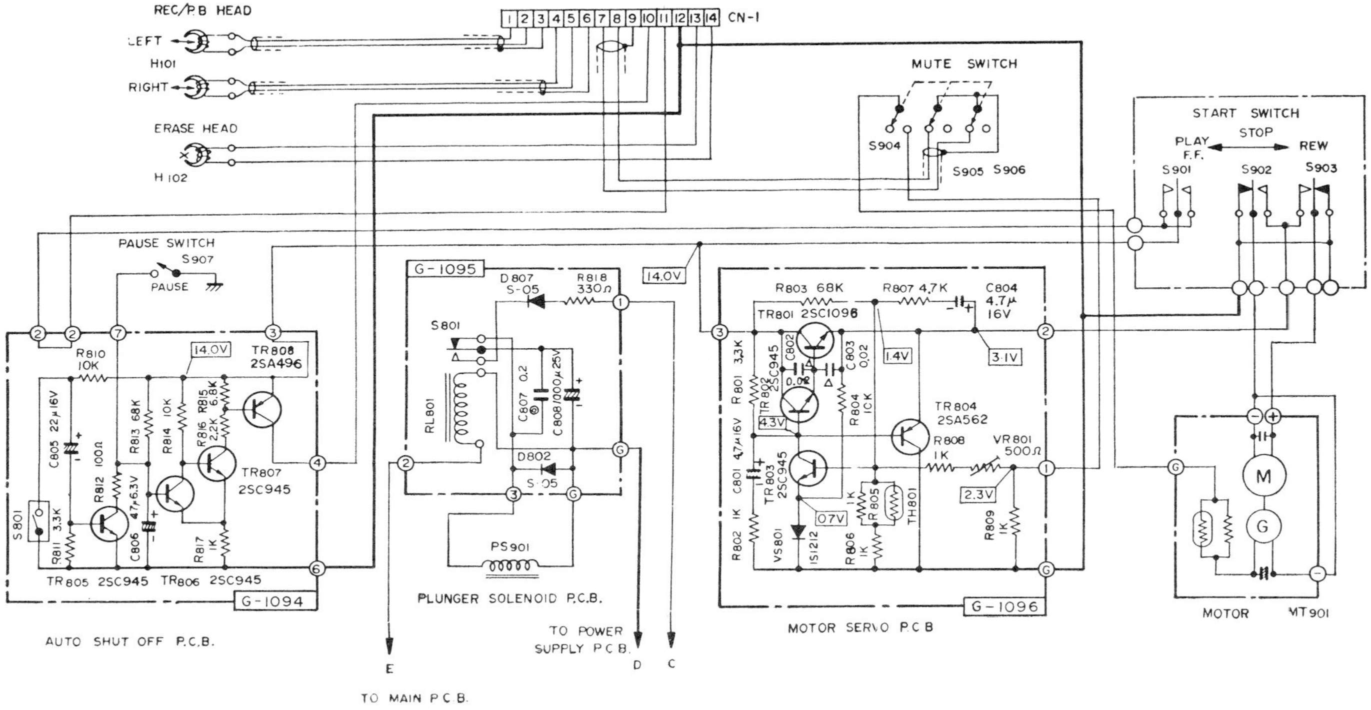

Fig. 18-2. Electronic circuits that control the motor and automatic shutoff system for the Sansui SC-700 stereo deck.

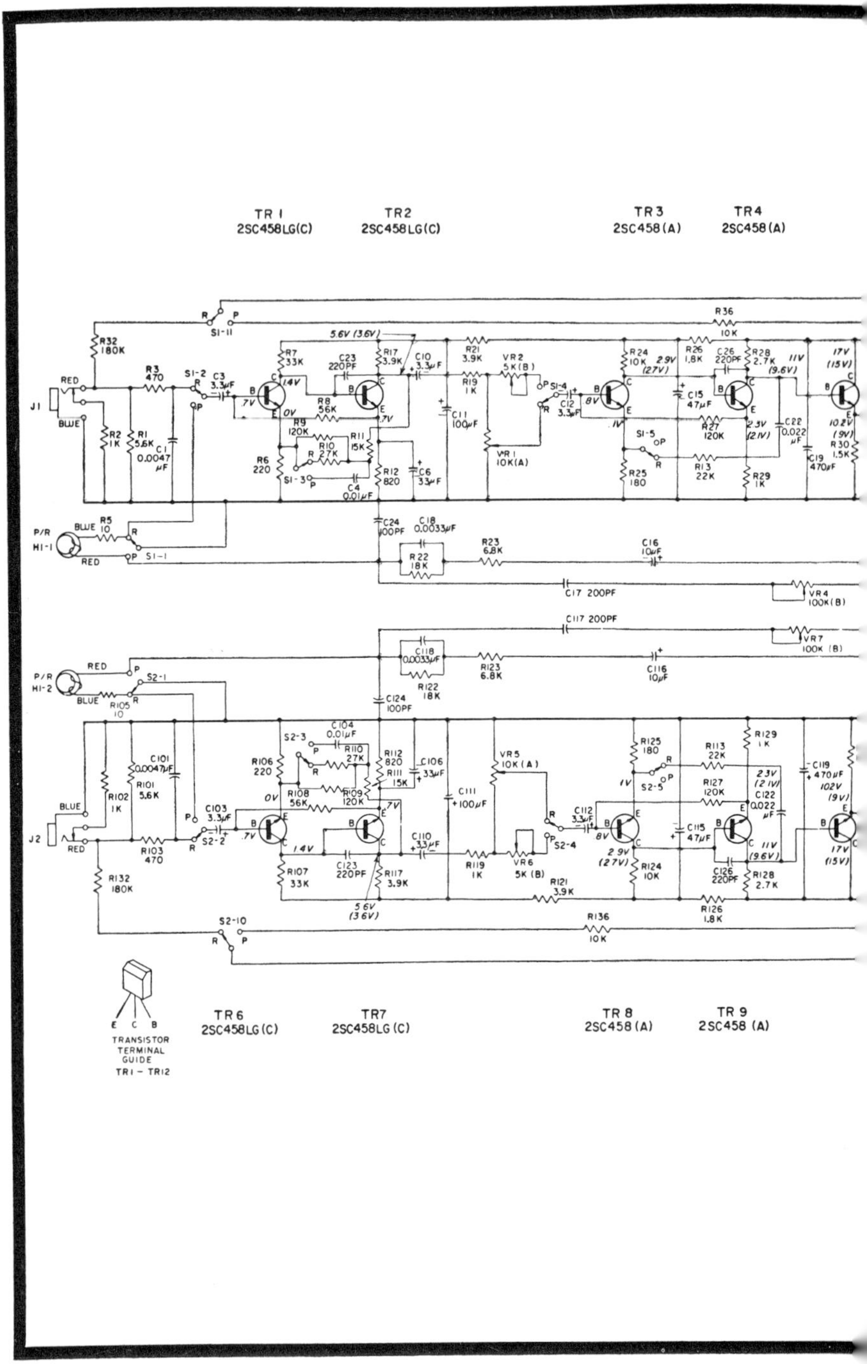
TR 1
2SC458LG(C)
TR2
2SC458LG(C)
TR3
2SC458(A)
TR4
2SC458(A)
TR6
2SC458LG(C)
TR7
2SC458LG(C)
TR 8
2SC458(A)
TR 9
2SC458 (A)
J1
J2
P/R
H1-1
P/R
H1-2
E C B
TRANSISTOR
TERMINAL
GUIDE
TR1 – TR12

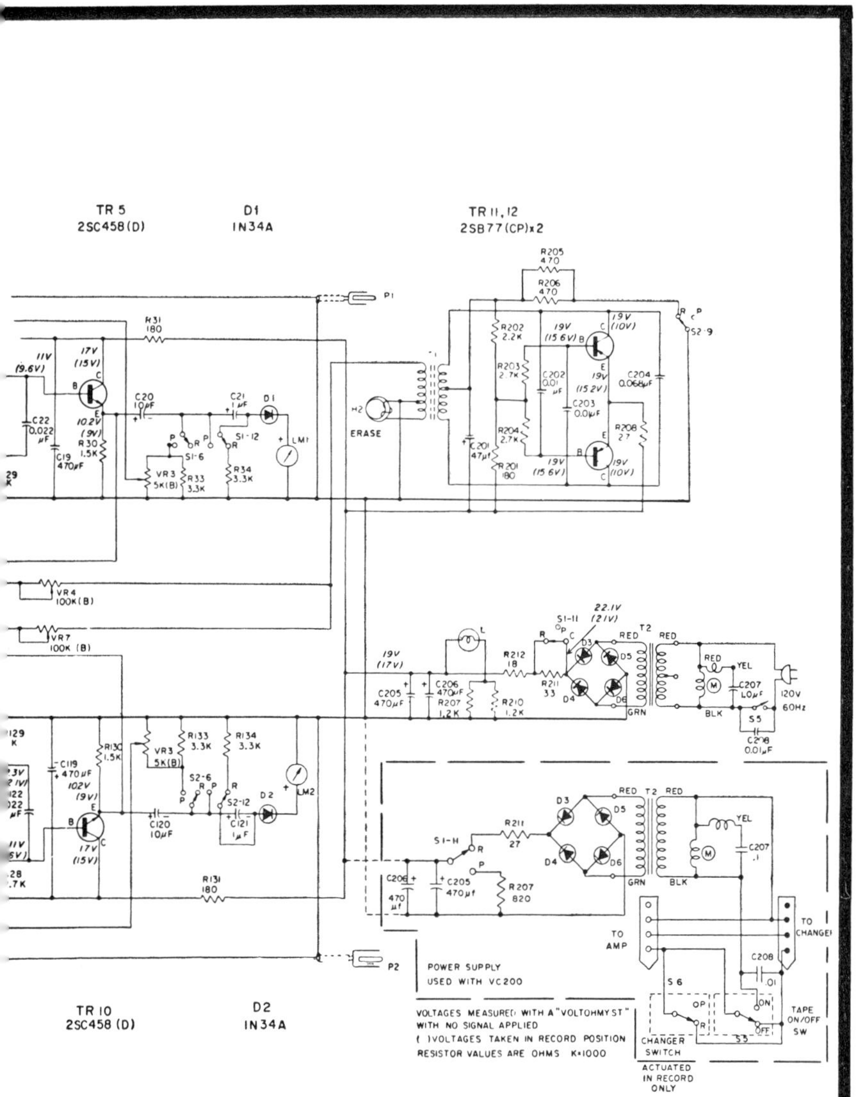

Fig. 18-3. Complete schematic of a typical stereo deck without switchable bias or noise-reduction circuitry. This is the RCA-MZD 563 and VC 200.

Belt Replacement

Belt replacement is an important servicing function that should be performed approximately every two years, whether or not the deck is in active use. This is because the rubber tends to crystallize and harden with age and may not provide adequate performance. This can be a key cause of wow and flutter in machines that are several years old.

Most belts can be replaced easily by loosening or removing two or three mounting screws that hold the capstan flywheel axle. The bearing clip on such a unit often can be moved far enough away from the flywheel thrust bearing so the belt can be slipped through. In some machines it may be necessary to remove the bearing clip entirely. When replacing the belt, be certain that no lubricating or other foreign material touches it as you slip it over the bearing shaft.

Some tape recorders require a very special kind of belt. TEAC decks, for example, call for a drive belt that has both a dull and a shiny side. The natural inclination would be to put the dull side next to the pulley, but TEAC's manual points out that the shiny side must go next to the pulley and flywheel; otherwise, serious wow and flutter will result.

Lubrication

Proper lubrication of the mechanical parts of the stereo deck can be somewhat more complicated than they are for a battery portable. The contact surfaces of moving metal parts should be lubricated with a light grease or Lubriplate compound. A drop or two of light machine oil should be applied to the capstan bushing assembly, turntable spindles, pressure-roller spindle, and winding drive-wheel spindle. If the machine you are servicing has additional rotating parts, each axle (spindle) should receive the same treatment—a drop or two of very light machine oil.

As for those fine contact points and the right kind of lubrication for them—different manufacturers recommend different types of grease. Philips calls for graphite grease, while Kenwood calls for molybdenum grease at these contact points. In fact, Kenwood provides a rather extensive breakdown of metal friction points that should by molybdenum-coated. They also call for a very light silicone oil in certain areas (see Fig. 18-5). Generally, lubrication is a matter of

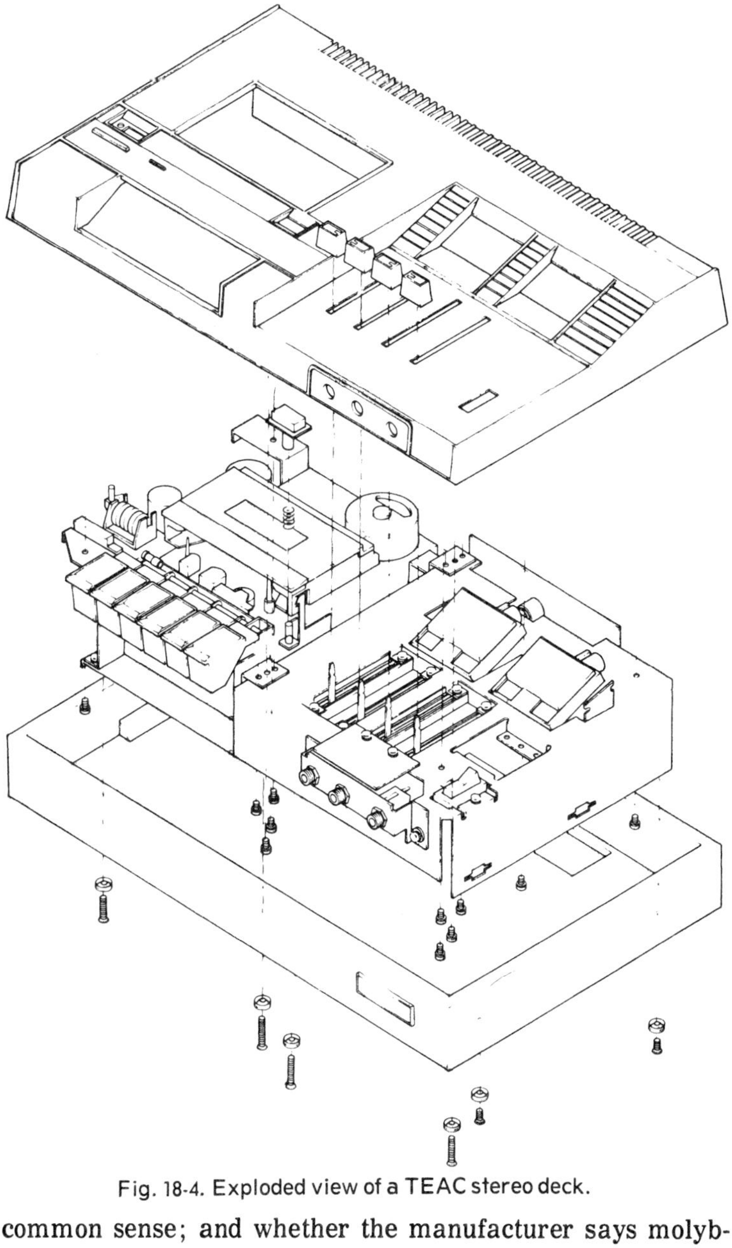

Fig. 18-4. Exploded view of a TEAC stereo deck.

common sense; and whether the manufacturer says molybdenum grease or graphite grease or anything else, an all-purpose medium-viscosity, heat-resistant lubricant like Lubriplate (GC Electronics) will serve the purpose for all recorders.

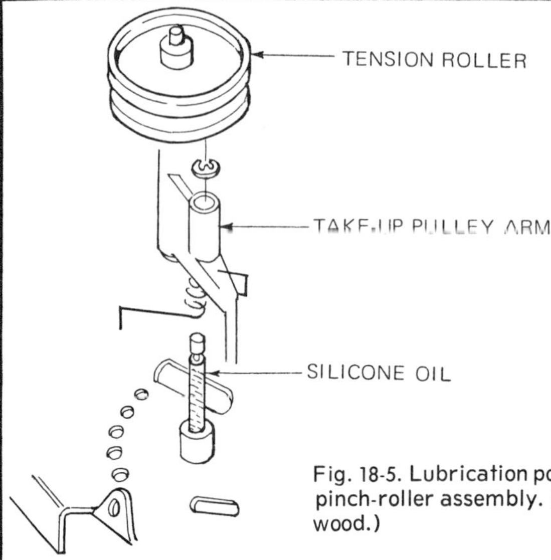

Fig. 18-5. Lubrication points in a typical pinch-roller assembly. (Courtesy Kenwood.)

Pinch-Roller Tension

Pinch-roller tension is an important adjustment on all cassette recorders. This was covered somewhat in the previous chapter for portable machines. For cassette decks this tension is even more crucial than for portables, and most machines provide some kind of adjustment for it. Generally, there is a spring with an adjusting screw or several notches or holes which permit a change in the spring tension. The required adjustment should be made according to the manufacturer's requirements, and can vary anywhere from 5 ounces to nearly 22 ounces. Going down the list (Fig. 18-6), Kenwood recommends a pinch-roller pressure of 10.56-15.84 ounces. TEAC recommends 5.23-7.04 ounces, SONY typically recommends 4.92-6 ounces, different Norelco units require pressures of 5.3-6.7 ounces, and 10-13 ounces, and a Realistic deck requires an incredible 17.2-21.5 ounces.

There are various ways to adjust this tension; the TEAC deck has a real Mickey-Mouse arrangement that requires you to bend the straight arm on the spring to increase the pinch-roller tension. And there are discrepencies. One of the best repair manuals available comes from Sansui, but nowhere is there any pressure adjustment shown. The only conclusion is that the pinch-roller pressure in the Sansui deck cannot be adjusted. Apparently, the spring must be replaced if there is a tension problem. The same is true for the Advent 401. Some typical methods of adjusting pinch-roller pressure are shown in Fig. 18-7.

Torque Measurements

Takeup torque is another problem area. A torque meter in the form of a torque spring scale or a special servicing cassette with a torque spring scale built in can be used to measure this. One such spring scale is advised by Sansui (see Fig. 18-8). Certainly, using this type of cassette spring scale is a lot easier than the open-type torque gages, although this means that both types must be in your tool kit. It certainly makes measuring takeup and rewind torque much easier.

To see what different manufacturers recommend for torque levels, see Fig. 18-9. Generally, this torque can be adjusted on the left reel hub spring in most decks. The alternative method of measuring reel torque is with a piece of tape attached to a hand-held torque gage clamped in a cassette hub as shown in Fig. 18-10. This method is obviously very cumbersome and is much more difficult than using the torque gage built into a cassette.

Flutter & Speed Measurement

Flutter and speed checks are a necessarily important measurement to make on all cassette recorders. Generally,

	GRAMS	OUNCES
NORELCO 2400	150-190	5.3-6.7
NORELCO 2401 (CHANGER)	280-370	10-13
KENWOOD KX-7010	300-450	10.56-15.84
SONY TC-165	140-170	4.92-6
TEAC A-21/22	300-400	10.05-12
TEAC A-350	150-200	5.28-7.04
REALISTIC SCT-3B	490-610	17.2-21.5

Fig. 18-6. Recommended pinch-roller tensions for some decks. Some manufacturers do not supply this information in their service manuals.

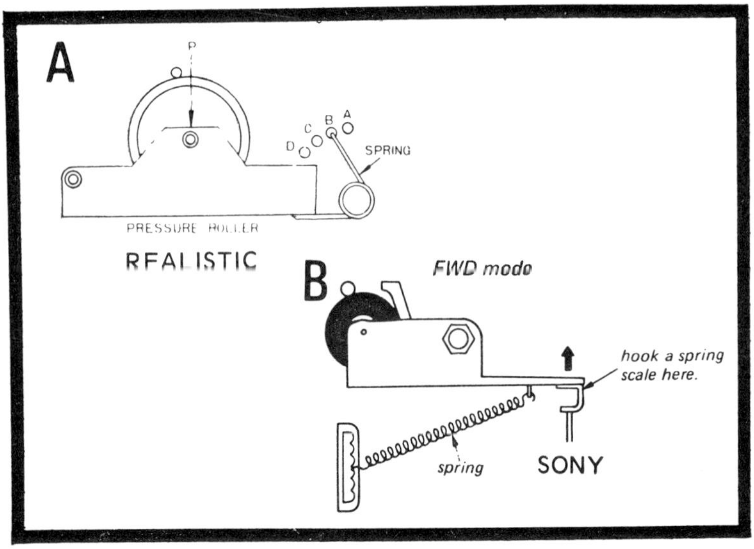

Fig. 18-7. Methods of adjusting pinch-roller tension in a Realistic deck (A) and a Sony deck (B).

these checks should be done only after the machine has received a complete cleaning, demagnetization, and lubrication. The setup requires attaching a flutter meter to one of the line output jacks. A digital frequency counter should be attached to the other line output jack. A flutter test tape cassette should be used; there are any number of these available. Some manufacturers have such test tapes as part of their equipment lineup. The signal used to make this check is a 3000 Hz tone, which most of these tapes contain.

If this test shows that the speed is off, it can be adjusted electrically in recorders that have an electric or electronic motor control. For some recorders, it's simply not possible to adjust the speed, and an entirely new drive train is called for. Where a serious off-speed condition exists and cannot be corrected, first try replacing the drive belt. The second step is to replace the flywheel, and the third step is to replace the motor. This last is a drastic measure and should be done only if it is obvious that the motor is off speed.

Flutter can be caused by a defective drive belt or simply by slippery contact surfaces on the motor pulley or on the flywheel itself. Clean both surfaces with a pencil eraser and try again. If this doesn't work, an application of nonslip compound (GC Electronics) may do the trick. This nonslip

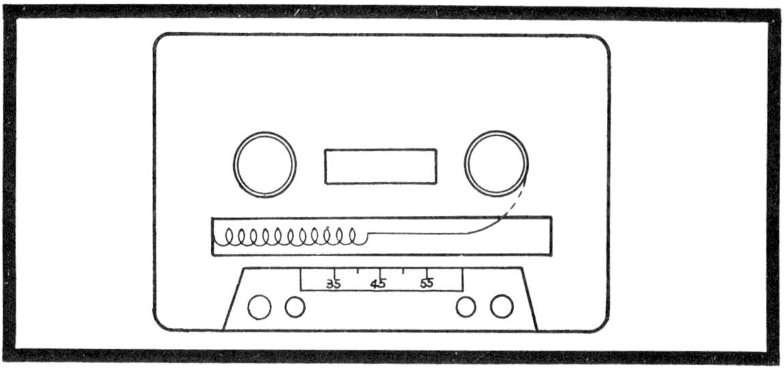

Fig. 18-8. Drawing of a torque-spring gage cassette from Sansui.

	TAKEUP	FAST FORWARD	REWIND
KENWOOD KX-7010	50-60	70-130	70-130
REALISTIC SCT-3B	40-70	60 or more	50 or more
SANSUI SC-700	35-45	35-45	40-50
SONY TC-165	55-75	80-100	80-100
TEAC A-21/22	45-60	80-120	80-120
TEAC A-350	40-60	70-150	70-150

Fig. 18-9. Some typical hub torques (in gram-centimeters) show the enormous variety recommended for different makes of decks.

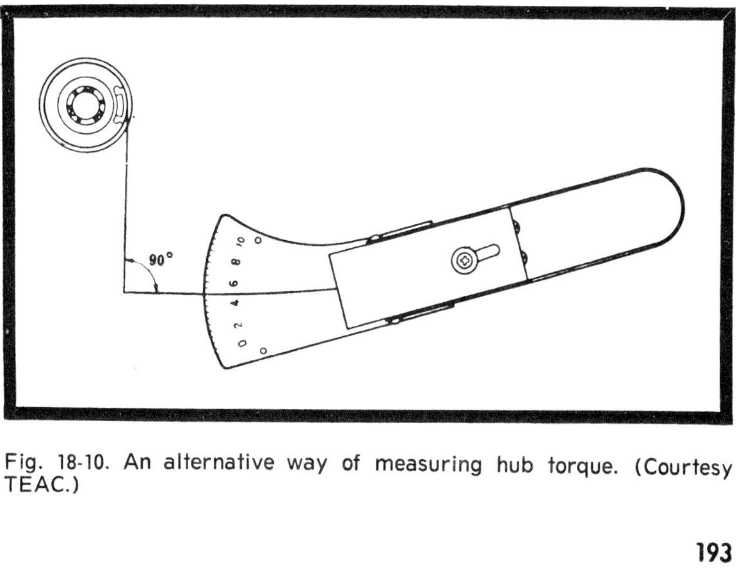

Fig. 18-10. An alternative way of measuring hub torque. (Courtesy TEAC.)

compound is delicate stuff and should be used very sparingly and carefully, and only as a last resort to prevent replacing the flywheel and motor.

Another possible cause of flutter is improper pinch-roller tension or a pinch-roller that has developed flat surfaces. If you suspect that a pinch-roller does indeed have "flats" or has become hardened with age, then you must replace the entire pinch-roller assembly.

ELECTRONIC MEASUREMENTS

Electronic tests, measurements, and calibration vary widely from one cassette deck to another, and it's a good idea to check the manufacturer's service manual before going ahead. Such tests include VU meter calibration, bias oscillator voltage, bias trap adjustment, recording level, record noise and erase measurements, chromium dioxide bias, equalization, and Dolby circuit testing and calibration. All of these can be pretty tricky, and all of them vary enormously with different types of units.

In all similar situations, especially where there seem to be electronic problems, be sure to check all appropriate switch contacts. A spritz of contact cleaner on the switch contacts can often work wonders. But even a simple task like this can lead you a merry chase. Some cassette decks have a switching arrangement that looks like a fugitive from a Rube Goldberg nightmare. The Advent Model 201, for example, instead of using one large complex record-playback switch, has a half dozen or so slide switches spotted around the printed-circuit board. Each one of these fits neatly into a slot in one large switch linkage plate. This is a metal plate with a lot of little cut-out slots. It moves the appropriate switches when the button is pushed at the top of the machine. While this makes individual function switch repair and replacement a lot easier, it also can make it somewhat more complicated to track down switch-related problems. Still, the cost of replacement parts is bound to be lower than for replacing one enormous complex function switch.

Most cassette deck troubleshooting is straightforward and common-sense type of work. It is the introduction of complicated and sophisticated circuits that make servicing such units a real challenge today. Some of these servicing problems are covered in the next chapter.

SPACE AGE CIRCUITS

Chapter 19

Two very important features are found on many of today's stereo cassette decks: Dolby noise-reduction circuits, and switchable bias and equalization circuits. In both cases, some adjustment is almost always called for when the machine is on the bench; sometimes repairs might be needed, too.

The adjustments—far from being simple—require either a thorough knowledge of the circuit or at the very least a service manual propped up on the bench. External test equipment called for depends on the manufacturer; some Dolby circuits can be adjusted very easily, while others take a more detailed setup.

BIAS ADJUSTMENTS

A typical cassette machine—the Advent 201—has a relatively simple adjustment for chromium dioxide tape bias. To set up, start with a Crolyn tape in the machine, the deck set in the record mode, the selector switch set for CrO2, and the noise reduction switched off. An external audio oscillator is used to feed a 1 kHz signal into the line input. The oscillator output should be adjusted to provide a 0 reading on the VU meter.

The next step is to reduce the input by 30 dB and to record a 1 kHz signal for two revolutions of the index counter. The external oscillator is then switched to 10 kHz and the higher frequency signal is recorded for two more revolutions of the digital counter. Then, rewind the tape cassette and play it back. If necessary, the Crolyn bias potentiometers (at the right rear of the printed-circuit board) should be adjusted for each channel. Then re-record the tape with the oscillator tone signals again, until the 10 kHz signal plays back at the same level as the 1 kHz signal.

One convenient aid provided for bias adjustment on this machine is to measure the bias voltage across the head terminals after removing the top protective plate. This is measured with a low-capacitance probe. Channel A is the bottom pair of terminals and channel B is the top pair. Adjusting the bias voltage approximately this way—by the number of dB alteration in the 10 kHz frequency response—will let you zero in on the proper bias setting. This is a trial-and-error type of adjustment; the nominal voltage is 14V for CrO_2 and 10V for regular ferric-oxide tape.

Since the same type of tape can vary in frequency response capability from one sample to another, you should always verify all bias adjustments by trying out another cassette of the same kind of tape. This is all that's required to set the bias on the Advent; a similar type of adjustment is provided on other machines.

The bias adjustment procedure for the Kenwood KX-7010, which uses only iron-oxide tape, is similar. In this case, the two pots involved in the bias circuit should be set approximately in the midway position as a starting point. The deck should be set in the record mode, and the oscillator coil itself should be adjusted so the output frequency is between 55 and 59 kHz. Then the bias trap coils must be adjusted so the bias current running through the head reaches its maximum level when terminated by a 10-ohm impedance.

The semi-adjustable variable resistors—the ones that are similar to the pots on the Advent unit—are 100K each. The bias current should be adjusted with these pots so it reaches 480 uA. This sets up the machine for standard (ferric oxide) tapes.

Adjusting the bias on the Sansui SC 700 requires an audio signal generator and an oscilloscope. The scope's vertical input terminal should be connected to the appropriate test points on the recorder's circuit board (Fig. 19-1). The audio generator must supply 105 kHz at 1.2V, which goes to the scope's horizontal input terminal. The selector on the scope should be set at "external." The recorder should be operated in the record mode with the pause button depressed. The first adjustment step is to touch up T501 for a Lissajous pattern as shown in Fig. 19-2.

Next, connect a frequency counter between terminals 13 and 14 of CN-1 (Fig. 19-1). Push the pause lever and set the unit in the record mode. Adjust T501 for a reading of exactly 105 kHz on the frequency counter.

To adjust the bias trap, connect a VTVM between the collector of TR401 and ground for the left channel and between the collector of TR402 and ground for the right channel (see Fig. 19-1). Again, with the pause lever depressed and the unit in the record mode, adjust L403 (left channel) and L404 (right channel) for a minimum reading on the VTVM.

The recording bias adjustment procedure starts with the tape oxide selector switch set to "standard." The output level should be set to maximum, and a VTVM connected to the line output terminals for left and right channels. An external 1000 Hz 200 mV signal should be fed into the input terminals for both channels. With pause button down, and the unit set in the record mode, insert a standard (ferric oxide) cassette. Adjust the mike / line level control for a reading of 0 VU on the deck's meters. Make sure that the reading on the VTVM is 1.2V at this point. Then, change the 1000 Hz signal from 200 mV to 20 mV. Release the pause lever and record the signal on the cassette. Change the signal from 1000 Hz to 10 kHz at 20 mV and record. During the recording, adjust VR501 for the left

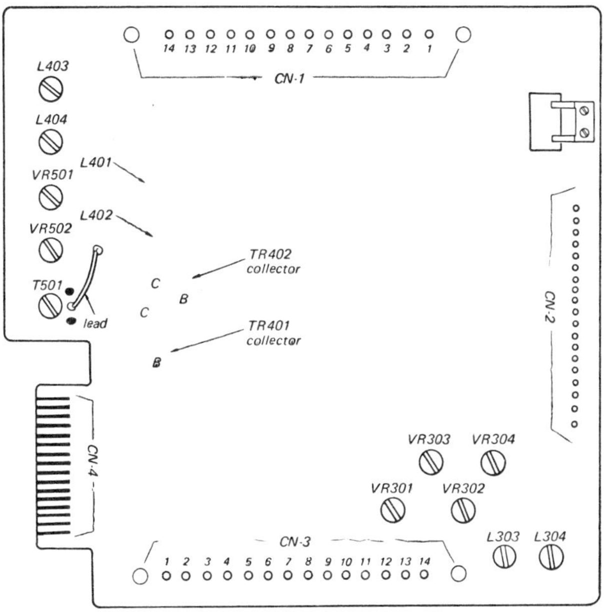

Fig. 19-1. Circuit board drawing showing reference points in the Sansui SC-700 where bias and Dolby adjustments are made.

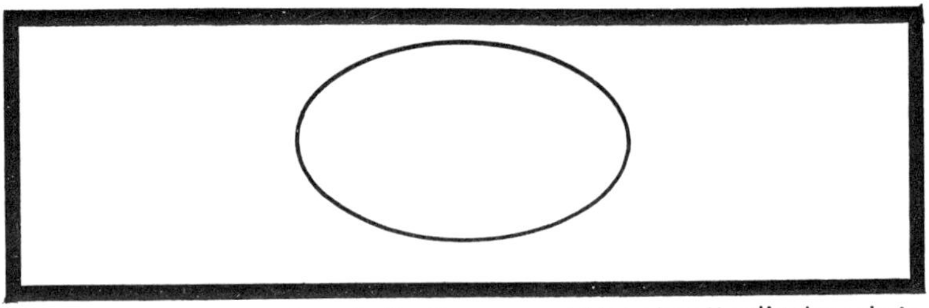

Fig. 19-2. Lissajous pattern indicating correct bias circuit adjustments in the Sansui SC-700.

channel and VR502 for the right channel so that the output of the 10 kHz playback signal is equal to that of the 1 kHz reference reading.

The next step is to set the bias for chromium dioxide tape. Set the output volume level to maximum, connect the VTVM to the left and right channel output terminals (high level), and supply a 1 kHz 200 mV signal to the input terminals for left and right. Then, with the pause lever pressed, set the selector switch for chromium dioxide and insert a chromium dioxide cassette. Adjust the mike/ line level for a reading of 0 V U on the machine's VU meters. Make sure that this reading on the VTVM is 1.2V.

Change the 1 kHz signal level from 200 mV to 20 mV, release the pause control, and record on the chromium tape. Then change the oscillator signal from 1 kHz to 12 kHz at 20 mV, and record. During recording, adjust VR005 for the left channel and VR006 for the right channel, so that the output of the 12 kHz playback signal is equal to the 1000 Hz reference reading. This step should be repeated several times for the greatest accuracy.

DOLBY ADJUSTMENTS

Adjusting the Dolby noise-reduction circuit requires another set of standards. In the Advent 201 deck, start with the machine in the record mode and the pause lever depressed; the noise-reduction switch should be off. Then, feed an external 5 kHz signal to the Channel A input. Adjust the external oscillator for 17.5 mV output at Channel A out on the printed-circuit board. Then move the probe to the Channel A1 test point and note the level (see Fig. 19-3).

Next, switch the noise-reduction circuit on and note an increase of 8 dB. Repeat this entire operation for channel B, using the B1 test point. If either channel needs alignment, turn the pot marked "law" fully counterclockwise, short the FET gate to ground, and adjust the gain pot for a 10 dB increase from the reference level at point A1 with the noise reduction switch off. Then, readjust the "law" pot with the noise reduction switch on for a 2 dB drop from the 10 level.

With the Sansui SC-700, the first step in adjusting the Dolby is to turn the Dolby switch off. Then it gets a little tricky. Unsolder the lead connected to oscillator transformer T501 (see Fig. 19-1). Turn VR001 and VR002 fully counterclockwise as viewed from the component side of the board. Next, connect TP001 and TP002 to ground. Connect a VTVM between terminal 12 or CN-3 (Fig. 19-1) and ground for the left channel and terminal 3 and ground for the right channel.

The input terminals for left and right channels should be fed a 5 kHz, 7 mV signal. Depress the pause lever and place the unit in the record mode. Adjust the mike/line volume level control for a reading of 17.5 mV.

Next, connect the VTVM between terminal 9 of CN-3 and ground for the left channel and between terminal 6 and ground for right channel. Check for a reading of 17.5 mV.

Now, turn the Dolby switch on and adjust VR003 for the left channel and VR004 for the right channel until the meter reads 10 dB higher than it did for 17.5 mV. Disconnect TP001 and TP002 from ground and adjust VR001 for the left channel and VR002 for the right channel until the meter reads 2 dB lower from 53.5 mV. Connect TP001 and TP002 to ground again. Make sure that the VTVM is connected between terminal 9 and ground for the left channel and terminal 6 and ground for right channel and reads 2 dB higher in this position. Resolder the wire lead to the board again.

If this sounds rather complicated, that's because the Dolby circuitry is just that; it's somewhat complex to adjust properly, and these adjustments are crucial. These two examples are representative of the many different machines that use Dolby circuitry; some are easier to set and calibrate than others.

Some adjustment may be needed for equalization for both recording and playback. The service manual is a must for all of these, just as it is for properly adjusting bias and Dolby circuits. Try to have a full-size schematic or pictorial in front of you when you're making these adjustments.

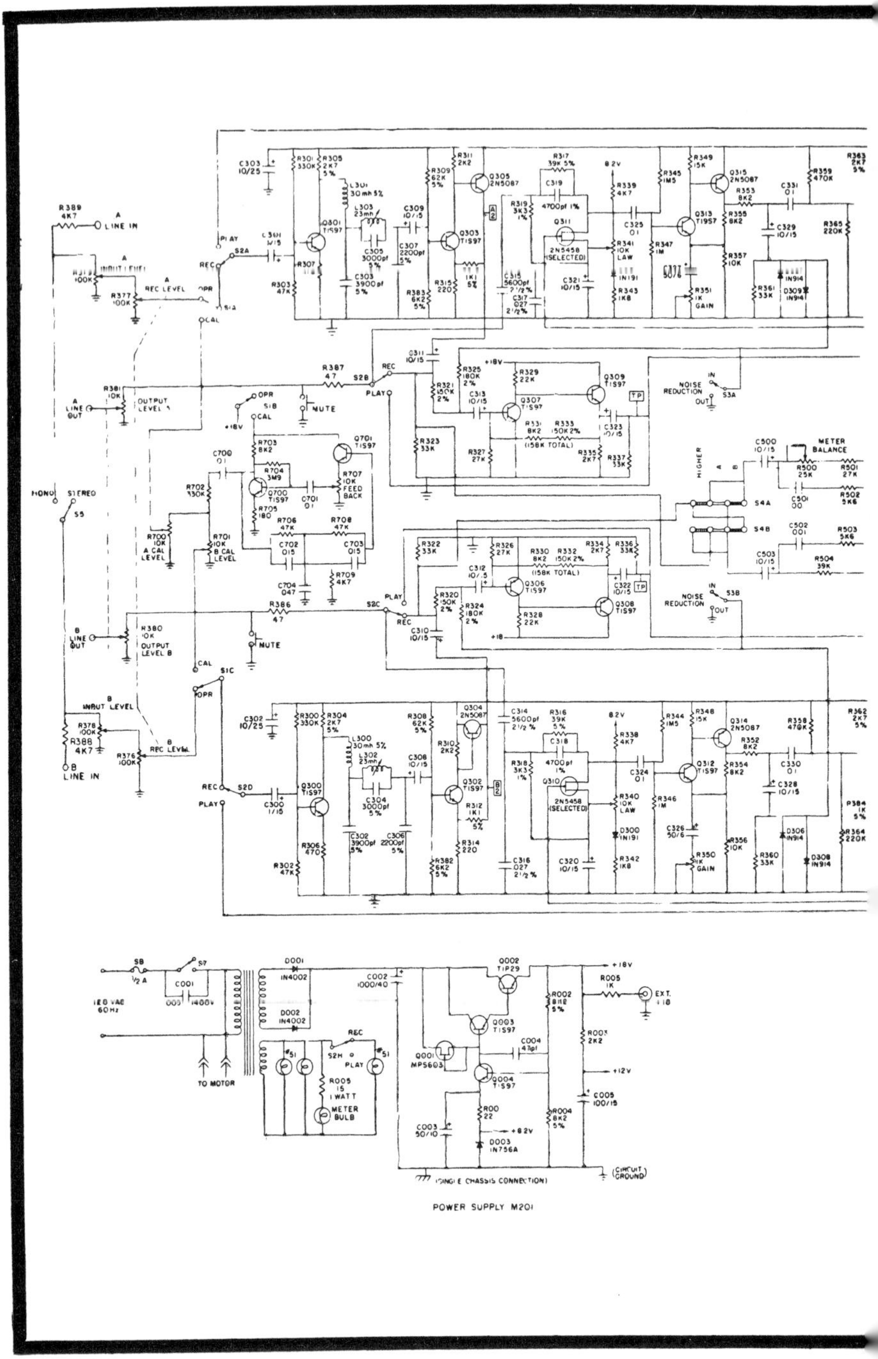
POWER SUPPLY M201

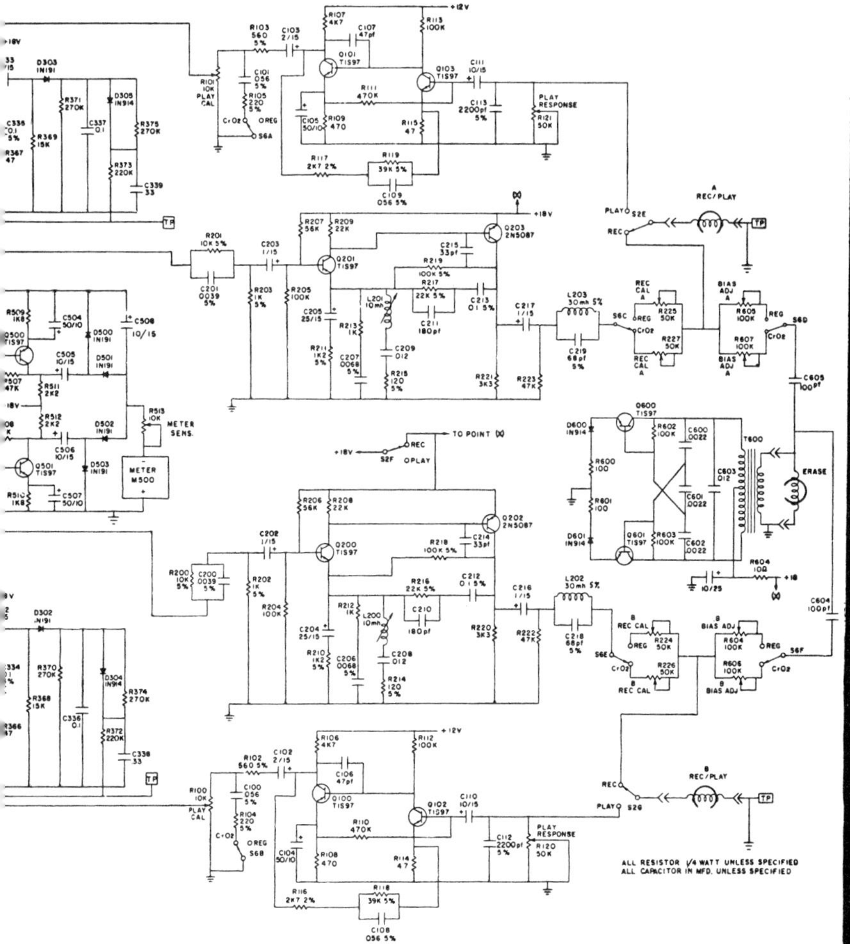

Fig. 19-3. The schematic of the Advent 201 includes crucial test points for Dolby and bias adjustments.

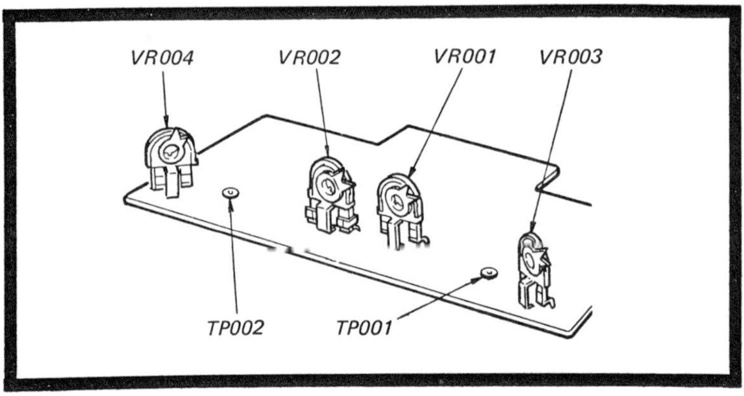

Fig. 19-4. Drawing of the bias-adjusting pots in the Sansui SC-700.

INDEX